INJECTION MOLDING REFERENCE GUIDE 4th ED.

This reference guide was originally prepared in 1990 as a convenient pocket sized resource for use in Injection Molding. This information is most useful by personnel who work in the injection molding field including press operators, technicians, engineers, designers, mold builders, etc.

There are many reference data tables regarding plastics data, statistical methods, engineering calculations and valuable training for personnel in the IM industry. The book includes basic part design, trig tables, calculations for thermal expansion, thermal exp coeffs, SHCS data, torque specs, shrink data, cooling time equation, mold debug guidelines, melt index data, resin density data, many tables of process guidelines, process development techniques, calculating heat load & water flow requirements, pipe data, conversion factors, transformer & motor current, PM & safety, basic statistics, equip selection guidelines and more.

Most books available on this subject are very informative, but require excessive reading time to capture the necessary information; this guide is designed to expedite learning and use thereof. This is accomplished via concise data tables, formulas, graphics as needed to convey the essential data and information.

Copyright & ISBN

A major effort has been made to achieve a highly accurate and informative book; any comments regarding improvements via corrections or new material to be included in future printings is solicited and welcomed.

The information in this booklet has been carefully prepared and is believed to be correct, but Advanced Process Engineering and Jay Carender make no warranties respecting it and disclaims any responsibility or liability of any kind for any loss or damage as a consequence of anyone's use or reliance upon such information.

Any reproduction of any part of this publication without the written permission of Advanced Process Engineering & Jay Carender is prohibited.

Copyright © 1990 by Advanced Process Engineering (Jay Carender) ... published as a pocket sized referenced booklet (3.5 inches wide x 7.5 inches tall with small fonts). Modified with improvements several times resulting in this, the 4th Ed in 1997.

Also re-formatted for epub and e-book distribution in 2011.

Contributions include miscellaneous figures/graphics used with permission from Conair Group, Franklin, PA. Contributions from RGJ, Inc., Traverse City, MI used with permission (Decoupled Molding^{SM} & Load Sensitivity).

Library of Congress, copyright number: TX0002998075

ISBN-13: 978-1466407824

ISBN-10: 1466407824

About the Author

The author - Jay Carender:
- Hands on processing skills,
- Mold build/mold design & engineering knowledge,
- Degree in Mechanical Engineering from GMI (General Motors Institute of Technology ... now known as Kettering University),
- DOE Training from Stat-Ease, Inc.,
- SPC Training from University of Tennessee Management Development Center,
- SPC Training from ASQC,
- Black Belt from AIT (Advanced Integrated Technologies),
- Productivity and Quality Improvement Training by Dr. Deming,
- Mold Design & Advanced Mold Design Training from New York University ... taught by John Klees Enterprises,
- Multiple courses on processing from RJG Industries, Inc.,
- Owner of Advanced Process Engineering - company started in 1990 to create and market pocket sized reference booklets for injection molding industry. Six booklets written & published along with other training manuals.

The hands on experience is from various fortune 500 companies performing injection molding. The extensive experience includes:
- Hot runner molding,
- High cavitation molds,
- High speed molding with cycle times less than 5 seconds,
- Stack molds, unscrewing molds, core pulls, slides, close tolerance parts,
- Engineering resins,
- SPC,
- Statistics,
- DOE to effect process improvement and dimensional nominalization.

Other APEBOOKS

Injection Molding Reference Guide, 4th Ed.
contains basic part design, trig tables, calc for thermal expansion w/ coeffs, SHCS data, torque specs, shrink data, cooling equation, mold debug guidelines, melt index data, resin density data, many tables of process guidelines, process development techniques, calculating heat load & water flow requirements, pipe data, conversion factors, transformer & motor current, PM & safety, basic statistics, equip selection guidelines and more.

Injection Molding Troubleshooting Guide, 2nd Ed.
contains troubleshooting tips/solutions for many injection molding defects, intro to DOE, discussion of VPT and Decoupled Molding^{SM} techniques (SM - RJG, Inc), and more.

Math Skills for Injection Molding, 2nd Ed.
contains intro to basic algebra, using conversion factors, percentages, ratios, proportions, trig as needed for draft and tapers, trig tables, thermal expansion calculations, calculate shrinkage, determining part cost, understanding efficiency and utilization, intensification ratios and clamp tonnage, projected area, residence time, cooling time, interpolation, heat load, Cp, Cpk, Pp, Ppk, correlation, math equations and samples for calculating piezo and strain gage transducer full scale pressure, and more.

Pocket Injection Mold Engineering Standards
mold spec sheets, quoting & design direction, shrinkage, mold steels and hardness, heat treatment, thermal conductivity, thermal expansion, plating, surface finish tables, cooling design guidelines, gate designs, runner sizing, venting, sprue pullers, sucker pins, ejection, slides, support pillars, alignment guidelines, O-ring guidelines, hot runner info, torque specs, trig tables and more.

Managing Variation for Injection Molding, 2nd Ed.
understand & quantify variation, 6 sigma techniques, Cpk, Ppk, Z-score math, correlation, single & multi regression analysis to create predictive equations, DOEs, ANOVA, components of variance - how to quantify % each, MSE & Gage R&R, SQC, control charts, real time SPC, process mapping, process qualification & validation, FMEA, nominalization, molding techniques to reduce variation, and more.

Basic Statistics and SPC
basic statistics for operators, inspectors, set-ups, etc to prepare personnel for more effective SPC techniques, includes training on understanding SPC control charts (variable control charts - Xbar & R; Xbar & MR and attribute control charts - p, np, c, u charts ... how to compute control limits). Also included: what is variation, cause & effect diagrams, root cause analysis, histograms, pareto analysis and Gage R&R. Discusses inherent problems w/ X-bar & R charts as used in injection molding; explains better sub-grouping strategy.

The Advanced Process Engineering Guide
a compilation of the first five APEBOOKS in one book ... coming soon!

For questions or comments, contact Jay Carender and Advanced Process Engineering at:

advproeng@aol.com
or
advproeng@gmail.com

Table of Contents

About the Author	2
Other APEBOOKS	3
Table of Contents (continued)	5
Basic Part Design (wall thickness & ribs)	6
Basic Part Design (scale-up issues, radii)	7
Basic Part Design (surface finish & draft angle)	8
Hardness Conversion Table	9
Thermal Conductivity & Specific Heat for Plastics	10
Thermal Conductivity For Selected Metals	11
Service Characteristics of Tool Steels	12
Surface Finish Comparison Table	13
Thermal Conductivity Graphs	14
Heat Transfer in Injection Molds	15
Cooling With Baffle Drops	16
Cooling With Bubblers	17
Vent Depths For Various Resins	18
Cold Runner Sizing	19
Calculating Thermal Expansion	20
Thermal Expansion Coefficients - Mold Materials	21
Torque Specifications For Fasteners	22
SHCS Dimensions (TYPICAL 1960 SERIES)	23
Shrinkage Values For Miscellaneous Resins	24
Calculating Shrinkage & Cavity Sizing	25
Common Abbreviations For Misc Resins	26
A Method For Estimating The Cooling Cycle	27
Molding Area Diagram (MAD)	28
New Mold Debug Checklist	29
Sample Mold Information Sheet	30
Sample Mold Process Sheet	31
Melt Index Test	32
Physical Properties Of Resins: ASTM Testing	33
Density Values - Selected Resins	34
Thermoplastic Identification	35
IR Spectroscopy	35
Melt Temperature Control and Measurement	36
Resin Process Temperatures	37
Calculate Heat Load and Cooling Required	38
Calculating Reynolds Number	39
API Schedule 40 & 80 Pipe Data	40
Basic Algebra	41
Conversion Factors	42
Basic Conversions	43
Temperature Convert °F to °C (001-200 °F)	44
Temperature Convert °F to °C (201-400 °F)	45
Temperature Convert °F to °C (401-600 °F)	46
Temperature Convert °F to °C (601-800 °F)	47
BAR to PSI Conversions	48
Calculate Area - Common Shapes	49
Millivolt Output For Type J T/C's (0 - 310 °F)	50
Millivolt Output For Type J T/C's (310 - 610 °F)	51
Ohm's Law For D.C. Circuits	52
Misc. Electrical Formulas	53
Full Load Transformer Current	54

Table of Contents (continued)

Full Load Motor Current	55
Elements Of A Molding Cycle	56
Calculating Plastic Machine Load	57
Hot Runner vs. Cold Runner	58
Hot Runner Startup & Operation	59
Injection Molding Press Selection	60
Checking Platens For Squareness	61
Injection Molding Press Preventive Maintenance	62
Injection Molding Safety	63
Screw Recovery vs Temperature	64
Injection Screw Terminology	65
Calculating Color Blend Ratios	66
Resin Drying Temperatures	67
Calculating Clamp Tonnage Required	68
Setting and Starting The Mold	69
Determining Gate Seal Time	70
Balance of Fill Analysis	71
Mechanics of Polymer Flow	72
Relative Viscosity Testing	73
Sample of Relative Viscosity Data	74
Viscosity Formula	75
Machine Load Sensitivity	76
Machine Load Sensitivity Data Example	77
Sources of Variation	78
Sample of Histogram Usage	79
Sample Pareto Chart Data	80
Examples of Process Capability	81
X Bar and R Control Charts	82
Example: X Bar and R Control Charts	83
Tables of Constants for Control Charts	84
Process Capability	85
Z Table for Process Capability Calculations	86
Standard Deviation, Mean & Normal Distribution	87
Tips, Tricks & Shortcuts	88
Tips, Tricks & Shortcuts (ASCII & Symbol Codes)	89
Equipment Selection (Molding Press)	90
Equipment Selection (Mold Size vs Tie Bar Space)	91
Equipment Selection (Tonnage & Injection Rate)	92
Equipment Selection (Plasticizing Rate)	93
Equipment Selection (Stroke & Daylight)	94
Auxiliary Molding Equipment - Dryer	95
Auxiliary Molding Equipment - MTCU	96
Auxiliary Molding Eq - MTCU (dump valves)	97
Auxiliary Molding Equipment - Color Feeder	98
Deming's 14 Points for Management	99
Decimal Equivalent of Fractions	100
Tap, Number & Letter Drill Sizes	101
Right Triangles - Find Angle	102
Right Triangles - Find Sides	103
Conversion Factors	104

Basic Part Design (wall thickness & ribs)

Resin Selection
Typically selected based on previous experience. These experiences are based on product performance requirements and economics. Performance issues require that we look at the resin's impact strength, flex modulus (indicator of stiffness), tensile strength, chemical resistance, weatherability, stress crack resistance or other physical properties needed by the molded product. Economics require that we review part price in terms of price per cu inch of part volume and processability characteristics. Remember that resin is purchased by the pound, and a low density resin requires less pounds for a given volume than does a high density resin. Polypropylene is the lowest solid density resin at 0.90 gr/cc (without foaming); thus, the widespread use of polypropylene. Other resins may have improved processability and more predictable shrinkage; thus, other selections.

Wall Thickness
- Attempt to have uniform wall thickness, but at least <u>avoid wall thicknesses which go from thin to thick</u>. Also, design gradual transitions in wall thickness.
- Remember – Doubling the wall thickness increases cooling time by approximately 400%. Resin supplier literature often provides typical L/T data (flow length vs. wall thickness).
- Wall, rib or boss intersections require good planning to avoid sinks. If a visible wall requires zero or minimal sink, then intersecting features below the visible wall should have wall thicknesses at only 50-60% of the visible wall. Some resins such as cellulosics, nylon and various filled resins can mold thick surfaces more easily without sinks; thus, allowing thicker walled ribs or bosses. Gate location and mold cooling also greatly affect resulting part appearance. With regards to ribs, a closer gate location helps with packing, but note also that the gate is best located so that fill direction is in line with ribs to enhance venting of the rib.

Shrinkage
The last resort for specifying shrinkage is the use of supplier data – this data may not be accurate enough for your application. If similar parts are currently molded from the same resin, then identify the mold dimension (accurately) and the part dimension, then shrinkage can be calculated. For very close tolerance parts, the new product should be prototyped in a unit mold to accurately identify shrink rates. Remember that shrink is affected by flow direction, packing pressure, fill rate and cooling rate; thus, the calculated shrinkage is only accurate if the process factors are duplicated in later processes and molds. Since shrink is affected by flow direction; do

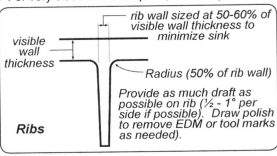

not apply the same shrink rate for all dimensions when the resin choice includes a high shrink crystalline resin such as polypropylene or polyethylene. For such crystalline resins the shrink rate is greater in the flow direction (approx 1/3 less for direction 90° to flow). Glass filled resins will typically have less shrinkage in the flow direction. There are also scale-up inaccuracies even when all factors are <u>apparently</u> duplicated. See list of potential scale-up errors on next page.

Basic Part Design (scale-up issues, radii)

Scale-up Inaccuracies

- Molder uses same hydraulic pressure, but with longer flow length resulting in more ΔP which results in less actual cavity pressure.
- Same hydraulic pressure, but mechanical advantage is less which results in reduced plastic pressure in mold; thus, reduced overall packing and greater shrinkage.
- Fill rate not duplicated. This writer has experience whereby molds fitted with pressure transducers indicated as much cavity pressure increase from 0.3 seconds faster fill time as is received from a 200 psi hydraulic pressure increase. Actual results were: 3285 psi difference from a 200 psi hydraulic psi increase (16.42:1 mechanical advantage), and 3200 psi difference caused by 0.3 second faster fill time. The total cavity pressure difference was 7159 caused by the pack and fill mentioned above plus minor effects by mold temperature changes (this data was obtained via a 2^5 multi variable type designed experiment .... 2^5 = 5 factors at 2 levels). See also the *IM Troubleshooting Guide, 2nd Ed.* to learn about press performance comparisons to verify that development process conditions can be duplicated in the production press.
- Same coolant temperature, but more cores and cavities in series resulting in a larger ΔT (temperature rise). This may also cause a reduced coolant GPM flow rate. If the temperature changes and/or the flow rate changes; the cooling rate changes. Record flow rates and ΔT in attempt to duplicate cooling. Larger molds will likely require more pressure to achieve flow rates similar to a smaller one cavity mold.
- Imbalanced fills can affect the process that each cavity sees; there is no imbalanced fill with a single cavity mold; some amount of imbalanced fill will exist in a multi cavity mold – this is inevitable; the key is to minimize the amount of imbalance.

Radii

- Adding radii to inside corners is needed for optimum part strength. Avoid sharp corners: resins which exhibit good notched izod impact strength properties should have inside corners radiused at a minimum of 50% of the wall thickness. More notch sensitive resins should receive a minimum radii up to 75% of the wall thickness.
- When possible, design corners with uniform wall (i.e. the outside surface follows the inside surface). This can be accomplished by the following radii relationships: inside radius @ 50% of wall and outside radius @ 150% of wall.

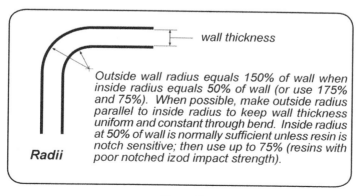

Radii — Outside wall radius equals 150% of wall when inside radius equals 50% of wall (or use 175% and 75%). When possible, make outside radius parallel to inside radius to keep wall thickness uniform and constant through bend. Inside radius at 50% of wall is normally sufficient unless resin is notch sensitive; then use up to 75% (resins with poor notched izod impact strength).

Basic Part Design (surface finish & draft angle)

Surface Finish
- Textured surfaces must be such that release from mold is possible. A rule of thumb often directs designers to provide 1° of draft per side for mold release (when possible); an additional 1° of draft may be needed for each 0.001" depth of texture (resin dependent).
- Softer resins with texture may show part scratches more easily as the molded peaks of the texture break off in part handling.
- Some resins release from a high polish with more difficulty. A good analogy is trying to slide Saran wrap over glass versus wood. A slightly abraded surface may release better than a high polish (resin dependent).

Draft Angles
Draft angles were mentioned previously relative to textures. As stated, a 1° draft per side enhances release and resulting processability. One degree draft angles are sometimes rejected by designers due to part fit requirements with mating components. In such cases, a compromise is needed allowing as much draft as possible whether it is ¼° or ½° per side. Shrinkage will be toward the core and away from the cavity; thus, lesser draft angles may be possible on surfaces which shrink away from the mold. In some cases where no draft or low draft is required other mold designs may be required such as changes to part orientation relative to parting line, parting line splits, lifters, more complex ejection mechanisms and/or special high lubricity coatings to the mold. Part designers should not use this previous statement as a blanket solution to difficult part designs, since such special mold engineering features complicate the mold and adversely affect mold cost. Parts with generous draft angles will typically permit faster cycles than those with marginal draft angles.

> Draft angles are often limited by mating part assembly and functionality requirements. Low or even zero draft angles are possible, but may affect the cycle and quality of the molded part.
>
> Provide as much draft as possible. Cavity half draft is important for part release from the cavity. Ejector half draft is needed to permit ejection without deformation. Good draft angles may permit faster cycles and improved part cosmetics.
>
> **Draft Angles**

Gate Location
- When possible, a single gate is preferred over multiple gates assuming a single gate will fill part with acceptable pressures. Multiple gates result in more weld lines and potential gas traps, increased mold complexity, increased gate vestiges and increased difficulty in balancing flow. Many larger parts will however require multiple gates to properly fill the part.
- Position gate to minimize the maximum flow length – locate gate central to different flow directions. Deviation from this rule may be needed to prevent flow hesitation. If flowing melt fronts simultaneously reach a thinner section and a normal thicker wall section: the flow may hesitate at the thin wall while flow proceeds down the path of least resistance.
- Locate gate to control weld lines so that the gate and resulting weld line is positioned away from stressed locations when part is in use. Note: Some gate locations result in no weld lines (e.g. top center gating on round closures, etc).

Hardness Conversion Table

BRINELL	ROCKWELL B 100 KG. LOAD	ROCKWELL C 150 KG. LOAD	TENSILE STRENGTH (1000 PSI) APPROX.
745		65.3	
712		—	
682		61.7	
653		60	
627		58.7	
601		57.3	
578		56	
555		54.7	298
534		53.5	288
514		52.1	274
495		51.6	269
477		50.3	258
461		48.8	244
444		47.2	231
429		45.7	219
415		44.5	212
401		43.1	202
388		41.8	193
375		40.4	184
363		39.1	177
352	(110)	37.9	171
341	(109)	36.6	164
331	(108.5)	35.5	159
321	(108)	34.3	154
311	(107.5)	33.1	149
302	(107)	32.1	146
293	(106)	30.9	141
285	(105.5)	29.9	138
277	(104.5)	28.8	134
269	(104)	27.6	130
262	103	26.6	127
255	102	25.4	123
248	101	24.2	120
241	100	22.8	116
235	99	21.7	114
229	98.2	20.5	111
223	97.3	(18.8)	108
217	96.4	(17.5)	105
212	95.5	(16.0)	102
207	94.6	(15.2)	100
201	93.8	(13.8)	98
197	92.8	(12.7)	95
192	91.9	(11.5)	93
187	90.7	(10.0)	90
183	90	(9.0)	89
179	89	(8.0)	87
174	87.8	(6.4)	85
170	86.8	(5.4)	83
167	86	(4.4)	81
163	85	(3.3)	79
156	82.9	(0.9)	76
149	80.8		73
143	78.7		71
137	76.4		67
131	74		65
126	72		63
121	69.8		60
116	67.6		58
111	65.7		56

VALUES IN ( ) ARE BEYOND NORMAL RANGE

Thermal Conductivity & Specific Heat for Plastics

RESIN	(THERMAL COND.) Btu/hr ft °F	(SPEC HEAT) Btu/lb °F
ABS	0.108 - 0.192	0.490
ABS/PC	0.142 - 0.217	
CA	0.100 - 0.192	0.410
CAB	0.100 - 0.192	0.390
HIPS	0.024 - 0.073	0.501
IONOMER	0.142 - 0.142	0.645
LDPE	0.192 - 0.192	0.760
MDPE	0.192 - 0.242	0.765
HDPE	0.267 - 0.300	0.870
PA 11	0.167 - 0.167	0.583
PA 12 GF	0.092 - 0.092	
PA 6 GF	0.173 - 0.173	0.669
PA 6	0.133 - 0.133	0.731
PA 6/6	0.142 - 0.142	0.735
PA 6/10	0.125 - 0.125	
PC	0.108 - 0.108	0.438
PC GF 10%	0.117 - 0.117	0.427
PC GF 40%	0.117 - 0.125	0.348
PET, PBT	0.167 - 0.167	0.490
PMMA	0.092 - 0.142	0.454
PMMA IMP	0.100 - 0.125	
POM	0.133 - 0.133	0.715
PP	0.066 - 0.079	0.667
PP COPOL	0.048 - 0.100	0.640
PPO	0.111 - 0.111	
PPO GF	0.092 - 0.092	
PPO/PS	0.125 - 0.125	0.519
GF PPS	0.167 - 0.259	0.497
PS G.P.	0.058 - 0.080	0.480
PSU	0.150 - 0.150	0.520
PVC	0.083 - 0.083	0.383
PVC RIGID	0.092 - 0.092	0.340
SAN	0.070 - 0.070	0.471
TFE	0.142 - 0.142	
TPU	0.041 - 0.041	

SPECIFIC HEAT & HEAT TRANSFER CONVERSIONS

1 Cal/sec cm °C = 2903 BTU-in/hr ft^2 °F
1 Cal/sec cm °C = 241.9 BTU/hr ft °F
1 W/(m °K) = 0.0023884 Cal/sec cm °C
1 W/(m °K) = 0.5778 BTU/hr ft °F
1 W/(m °K) = 6.9335 BTU-in/hr ft^2 °F
1 Btu/(Lb °F) = 4.184 KJ/(Kg °K)
1 Btu/(Lb °F) = 4184 J/(Kg °K)
1 Cal/(g °C) = 1 Btu/(Lb °F)

Thermal Conductivity For Selected Metals

METAL	Btu/hr ft °F
ALUMINUM ALLOYS:	
1100	137.88
2024	108.86
6061	99.18
7075	70.15
COPPER AND ALLOYS:	
Pure copper	227.63
Free machining (1%Pb)	222.55
Cartridge brass (70%)	70.15
Naval brass	67.73
Manganese bronze	62.89
Phosphor bronze	41.12
Ampco 18 @ 92 R_b	36
Ampco 21W @ 29 R_c	25
Ampco 22W @ 35 R_c	25
Ampco 940 @ 94 R_b	125
Ampco 97 @ 77 R_b	190
BeCu Moldmax® @ 40 R_c	60.48
BeCu Moldmax® @ 30 R_c	75.47
BeCu Protherm® @ 96 R_b	145.62
(BeCu alloys from Brush-Wellman)	
IRON:	
Pure iron	43.06
Cast iron	27.09
Low carbon steel alloys	30.00
High carbon steel alloys	26.13
TOOL STEEL:	
P-20	22.01
H-13	18.63
S-7	16.45
M-2	12.34
T-15	12.10
D-2	11.37
STAINLESS STEELS:	
303 SS	9.68
410 SS	14.51
414 SS	14.27
420 SS	14.27
440 SS	13.79
TITANIUM ALLOY	9.19
INVAR (36% Ni)	6.05

Service Characteristics of Tool Steels

AISI NO.	THERM CONDUCT. (Btu/hr ft °F)	HARDNESS (Rc)	TEMPERING TEMP (°F)	DISTORTION DURING HARDENING	COMPRESSIVE STRENGTH	TOUGHNESS	RESISTANCE TO WEAR
A2	15	60/62	400	LOWEST	VERY HIGH	MEDIUM	HIGH
A6	15	58/60	350	LOWEST	HIGH	MEDIUM	MED - HIGH
A10		58/60	400	LOWEST	VERY HIGH	MEDIUM	HIGH
A11		60/62	1000	LOW	HIGH	LOW	HIGHEST+
2% BeCu[1]	12.4	37/41	600	PREHARDENED	MED	LOW	VERY LOW
0.5% BeCu[2]	60.5	17/19	600	PREHARDENED	LOW	LOW	VERY LOW
D2	145.6	56/58	700	LOWEST	HIGH	LOW	HIGH - VERY HIGH
H13	11.4	50/52	1050	MED - HIGH	MED - HIGH	VERY HIGH	MED
L6	18.7	58/62	500	LOW	HIGH	VERY HIGH	LOW - MED
M2		60/64	1100	LOW MED	HIGHEST	LOW	VERY HIGH
O1	12.3	58/60	500	VERY LOW	VERY HIGH	MED	MED
O6	18.5	58/60	500	LOW	VERY HIGH	MED	HIGH
P20[3]	18.5	28/34	1150	PREHARDENED	MED	VERY HIGH	LOW
S7	22	54/56	550	LOWEST	HIGH	VERY HIGH	LOW - MED
S7	16.5	50/52	1000	LOWEST	HIGH	VERY HIGH	LOW - MED
420 SS	16.5	50/52	850	LOW	MED - HIGH	MED	MED
440 SS	14.3	56/58	425	LOW	HIGH	LOW	HIGH

[1] A.K.A. MOLDMAX® FROM BRUSH-WELLMAN OR SOMETIMES REFERRED TO AS ALLOY 25
[2] A.K.A. PROTHERM® FROM BRUSH-WELLMAN
[3] TYPICALLY USED FOR PROTOTYPE MOLDS OR LOWER COST, SHORT RUN PRODUCTION MOLDS

Surface Finish Comparison Table

RMS Micrometre (µM)	RMS Microinch (µIN)	Charmilles (LOG #)	SPI FINISH OLD	SPI FINISH CURRENT
.00 - .03	0 - 1		1	A1
.03 - .05	1 - 2		2	A3
.05 - .08	2 - 3			
0.1	4	0		
0.14	5	2		
0.16	6.4	4		
0.18	7.2	5	3	B3
0.2	8	6		
0.25	10	8		
0.28	11.2	9		C3
0.32	12.8	10	4	
0.4	16	12	(280 stone)	
0.45	18	13		
0.5	20	14		
0.56	22.4	15		
0.63	25.2	16		
0.7	28	17	5	D2
0.8	32	18	5	
0.9	36	19		
1	40	20		
1.12	44.8	21		
1.26	50.4	22		
1.4	56	23		
1.6	64	24		
1.8	72	25		
2	80	26		
2.2	88	27		
2.5	100	28		
2.8	112	29		
3.2	128	30		
3.5	140	31		
4	160	32	6	D3
4.5	180	33	6	
5	200	34		
5.6	224	35		
6.3	252	36		
7	280	37		
8	320	38		
9	360	39		
10	400	40		
11.2	448	41		

NOTE: RMS (root mean square) is approx 11% higher than Ra or AA - arithmatic average .... which is average of all areas above and below mean (determined by height and width of peaks and valleys).

Mold Surface Finish Terminology

	Grit	Media
A 1	8000+	# 3 diamond buff
A 2		# 6 diamond buff
A 3	1200+	# 15 diamond buff
B 1	# 600	paper
B 2	# 400	paper
B 3	# 320	paper
C 1	# 600	stone
C 2	# 400	stone
C 3	# 320	stone
D 1	# 11	glass bead (dry blast)
D 2	# 240	oxide (dry blast)
D 3	# 24	oxide (dry blast)

Thermal Conductivity Graphs

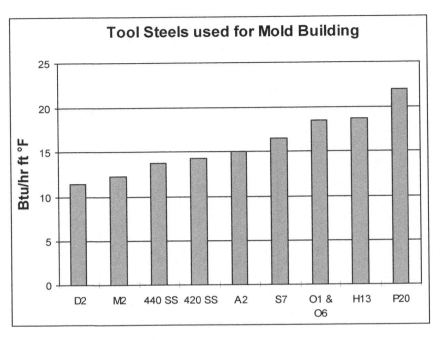

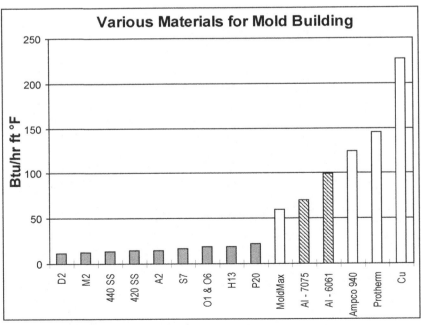

Heat Transfer in Injection Molds

There are large differences in thermal conductivity between different tooling materials. These differences are important to understand when planning injection mold construction to achieve optimum cycle times.

Five Main Variables Affecting Heat Transfer
1. Distance to coolant from part
2. Core/cav thermal conductivity
3. ΔT (water temp vs. plastic melt)
4. Amount of heat to remove
5. Total area core/cavity

Two Variables Which Mold Designer Can Control
1. Distance to coolant from part
2. Core/cav thermal conductivity

Algebraic Equation for Thermal Conductivity

$$\text{Time (hrs)} = \frac{L \times H}{K \times A \times \Delta T}$$

L = Distance (ft)   H = Heat (Btu)
A = Area (ft^2)   ΔT = Resin melt temp - coolant temp (°F)
K = Thermal conductivity (Btu/hr ft °F)

Drilled Coolant Lines

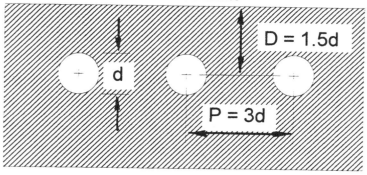

Typical Values for "D" (distance) & "P" (pitch)
D = 1.5 - 2 X d (diameter)
P = 3-5 X d (diameter)

Preferred Values for "D" (distance) & "P" (pitch)
D = 1.5 X d (diameter)
P = 3 X d (diameter)

Cooling With Baffle Drops

Baffle drops are typically designed whereby several drops are in series; thus, not recommended for small cores of substantial length. The pressure drop in such designs results in low flow rates with large temperature rise in the cooling circuit (bubblers are often the preferred design for many small cores, as each core is in parallel).

Formula for Equivalent Hyd Diameter

$$D_h = \frac{4A}{P}$$

$$A = \frac{\pi d^2}{4}$$

$$P = \pi d$$

A = area of non round hole
P = perimeter of non round hole

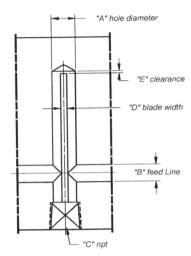

BAFFLE DROP SIZING

A		B	C				D	E
hole diam	(½ round) hyd diam	suggested feed line	pipe size	Travel per rev	Travel per ½ rev		blade width	gap
inches	inches	inches	NPT	inches	inches		inches	inches
0.250	0.116	0.250	1/16-27	0.037	0.0185		0.062	0.125
0.312	0.154	0.312	1/8-27	0.037	0.0185		0.062	0.156
0.438	0.212	0.438	1/4-18	0.056	0.028		0.094	0.219
0.562	0.289	0.438	3/8-18	0.056	0.028		0.094	0.281
0.688	0.367	0.438	1/2-14	0.071	0.036		0.094	0.344
0.938	0.502	0.562	3/4-14	0.071	0.036		0.125	0.469
1.125	0.617	0.688	1-11½	0.087	0.043		0.125	0.563

[1] Hydraulic diameter computed using formula above
[2] Needs to be at least equal or greater than hyd diameter.

Cooling With Bubblers

A bubbler consists of a delivery tube located inside of a drilled water passage. Bubblers are useful in cooling long thin cores where drilled passages do not run thru (i.e. blind hole). See figure below for typical layout. An annulus is the donut shaped area between two concentric circles (i.e. area of hole ID minus tube OD). There exists at least two methods (listed below) for sizing the optimum relationship between hole ID and tube OD for maximum coolant flow. The first formula approximates an equal area between tube ID and annulus. The second formula is based on hydraulic diameter, but results in different areas. When using equal areas for small bubblers the gap thickness of the annulus becomes quite narrow which may result in larger pressure drops and reduced flow. If the hyd diameter method is used, then the tube ID can become restrictive relative to the gap. It is therefore suggested that a spreadsheet table be constructed listing sizes using both formulas; the optimum may well be some weighted average of the two methods.

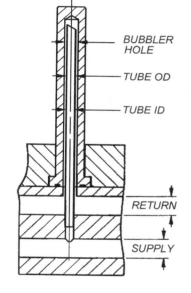

Bubbler hole = (tube ID + Tube wall) ÷ 0.707*
*closely approximates equal areas of tube ID and annulus

Bubbler hole = (tube id + tube OD)**
** based on equal hydraulic diameter of tube ID and annulus; hydraulic diameter of tube ID is tube ID, but hydraulic diameter of annulus must be calculated using same formula listed for baffle drops (non-round holes); formula applied to annulus reduces to the above relationship.

NOTES:
1. When bubbler feed tube ID is 0.062" or below, install a 50 micron (0.002") bag filter on supply side of circuit (or other suitable filter).
2. Core pins (OD) between 0.125" and approximately 0.156" should be cooled by air OR heat conducting copper rod with base surrounded by water (air is best with open ended core which may be telescoped during mold closed; use copper for blind holes). Bubbler circuits used in long thin cores (> 0.188" OD) should be designed in parallel to avoid the high pressure drop if designed in series. For bubblers in parallel, the coolant return passage is restricted by the bisecting bubbler passing thru this level; this may require the coolant return diameter to be larger. The coolant supply area should at least be equal to or greater than the sum of tube ID areas. Preferably, larger to allow for pressure drops.

Vent Depths For Various Resins

RESIN	DEPTH (inches)		
ABS	0.0010	-	0.0015
ACETAL	0.0005	-	0.0010
ACRYLIC	0.0015	-	0.0020
CELLULOSE ACET, CAB	0.0010	-	0.0015
ETHYLENE VINYL ACET.	0.0010	-	0.0015
IONOMER	0.0005	-	0.0010
NYLON	0.0003	-	0.0005
PPO/PS (NORYL)	0.0010	-	0.0020
POLYCARBONATE	0.0015	-	0.0025
PET, PBT, POLYESTERS	0.0005	-	0.0007
POLYSULFONE	0.0010	-	0.0020
POLYETHYLENE	0.0005	-	0.0012
POLYPROPYLENE	0.0005	-	0.0012
POLYSTYRENE	0.0007	-	0.0010
POLYSTYRENE (IMPACT)	0.0008	-	0.0012
PVC (RIGID)	0.0006	-	0.0010
PVC (FLEXIBLE)	0.0005	-	0.0007
POLYURETHANE	0.0004	-	0.0008
SAN	0.0010	-	0.0015
T/P ELASTOMER	0.0005	-	0.0007

NOTES:
1. Vent land is typically 0.04" - 0.12" long.
2. Vent relief is 0.03" deep & 0.25" - 0.5" wide.
3. Vent relief must run out to atmosphere.
4. Vacuum venting should only vent to vacuum source
5. Merits of vacuum venting are:
   a. Longer interval between cleaning.
   b. Can accomplish same venting with vent of reduced depth.
6. Vents should be ground in.
7. Vents & relief should be consistent throughout common mold components.
8. Deeper vents can be used with GF resins.
9. Runner system should also be vented.
10. High melt flow rate resins require vents of lesser depth.
11. Vents that are in a moving pin are self cleaning and preferred over non-moving pins.
12. Make vents as accessible as possible for cleaning purposes.

These are general guidelines; refer also to the supplier design manuals for additional information.

Cold Runner Sizing

Runner sizing should be done per the following:
A. Identify maximum part wall thickness.
B. Size runner by <u>working from part back to sprue</u>.
C. Set runner adjacent to gate at diameter equal to maximum part wall thickness.
D. Size each <u>upstream</u> branch by taking the number of branches (typically 2) to the 1/3 power X previous size. NOTE: $2^{1/3} = 1.259$
E. Repeat previous step until all branches sized.
F. If primary (largest) runner sizes appear too large and may lengthen the cycle, then oversized runners may be reduced in diameter by a selected percentage. A mold filling analysis can be done to identify fill pressures.
G. Remember to err on the small or "steel safe" side; it is easier to cut the runner bigger after the first mold trial than to re size smaller.

Number of Branch Runners	Ratio: Feeder to Branch Diameter
2	1.259
3	1.442
4	1.587
5	1.709
6	1.817
7	1.913
8	2.000

Cold runners should have either a round or modified trapezoid cross section. A round runner is cut with half the runner in each plate or mold half. A modified trapezoid is cut into one plate. It has a full radius at the bottom. The depth is equal to twice the radius. It has 5 degrees draft tangent to the radius on either side.

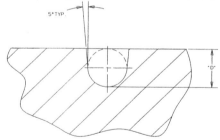

Calculating Thermal Expansion

The following formula is used to calculate thermal expansion:

$$\delta t = \alpha (\Delta T) L$$

Where the symbols mean the following:

δt is elongation due to temp change

α is the coefficient of thermal expansion (see data table on next page)

ΔT is temperature change .... ($T_{HIGH} - T_{LOW}$)

L is the length subjected to the expansion

Example:

The leader pins on a mold are 26 inches apart. The mold is heated from 70° F room temperature to 140° F on the stationary half only and the ejector half is cooled to 55° F; if the moldbase is steel, use 6.6 X 0.000001 as the coefficient of thermal expansion or use table if specific steel is known.

The expansion is calculated as follows:

$$\delta t = 6.6 \times 0.000001 \times (140 - 55) \times 26$$
$$\delta t = 0.0146 \text{ inches}$$

Note: 0.0146 inches is enough to cause mold closing problems due to misalignment of the leader pins.

Thermal Expansion Coefficients - Mold Materials

μ = micro....multiply X .000001; μ in/in from 68 °F to

TYPE	200 °F	400 °F	800 °F
1020	6.5	6.7	7.1
4140	6.8	7.1	
6150	6.8	7.1	7.4
W1	5.8	6.1	7.3
W2	7.4		8.0
S1	6.9	7.0	7.5
S5	6.4		7.0
S6	6.4		7.0
S7	6.8	7.0	7.4
O1	5.8	5.9	7.1
O2	6.2	7.0	7.7
A2	5.8	5.9	7.2
A6	6.4	6.9	7.5
D2	5.6	5.7	6.6
D3	6.3	6.5	7.2
D4	6.2		6.9
H10	6.1		6.8
H11	6.2	6.9	7.1
H13	5.8	6.4	6.8
H14	6.1		
H19	6.1	6.1	6.7
H21	6.9	7.0	7.2
H22	6.1		6.4
T1	5.3	5.4	6.2
T5	6.2		
T15		5.5	6.1
M1		5.9	6.3
M2	5.6	5.2	6.2
M3			6.4
M4			6.4
M7		5.3	6.4
M10			6.1
L2	7.4		8.0
L6	6.3	7.0	7.0
P2	7.0		7.6
P20	6.5		7.1
SS 303,4	9.6	9.9	
SS 316	8.8	9.0	
SS 414	5.8	6.1	
SS 420	5.7	6.0	
SS 440	5.7		
Ti alpha alloy	4.6		4.8
Ti alpha-beta	5.0	5.1	5.2
Ti beta alloy	5.2	5.4	5.6
Ampco 18, 21, 22	9.0		
Ampco 940	9.7		
FREE MACH Cu	9.8		
BeCu (2%)	9.7		
BeCu (0.5%)	9.8		
6061 Aluminum	13.1		
INCONEL	6.4		

Torque Specifications For Fasteners

GENERAL TORQUE SPECIFICATIONS
ENGLISH FASTENERS (FOOT-POUNDS)

MATERIAL GRADE / BOLT SIZE	SAE 2 MILD STEEL	SAE 5	SAE 8	SHCS	BRASS	SS AISI 303
1/4-20	6	11	12	13	5	5
1/4-28	7	13	15	16	6	7
5/16-18	13	21	25	27	8	9
5/16-24	14	23	30	33	9	10
3/8-16	23	38	50	52	15	17
3/8-24	26	40	60	60	16	18
7/16-14	37	55	85	86	23	25
7/16-20	41	60	95	95	25	28
1/2-13	57	85	125	130	32	37
1/2-20	64	95	140	145	34	40
9/16-12	80	125	175	180	44	50
9/16-18	91	140	195	210	48	54
5/8-11	111	175	245	255	68	75
5/8-18	128	210	270	290	73	80

GENERAL TORQUE SPECIFICATIONS
METRIC FASTENERS (NEWTON METERS)

MATERIAL CLASS / MM-DIAM	4.6	4.8	5.8	8.8	9.8	10.9	12.9
5	3	4	5	7	8	11	12
6	5	6	8	12.5	14	17	20
6.3	5.5	8	9.5	14	16	21	24
8	12	16	20	30	34	44	50
10	23	32	40	60	70	85	100
12	40	56	70	103	120	150	180
14	65	90	110	167	190	240	280
16	100	140	170	270	290	380	440
18	137	177	225	350	--	480	580
20	200	--	330	520	--	740	860

Note: check also torque recommendations from your fastener supplier and/or equipment/product manufacturer for item which fasteners are being used. These values are approximate. SHCS used for Husky, Moldmaster and other hot runner systems may use slightly lower torque values; see supplier guidelines.

SHCS Dimensions (TYPICAL 1960 SERIES)

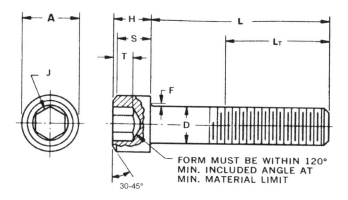

SIZE	D (MAX)	A (MAX)	H (MAX)	S (MIN)	J (NOM)	T (MIN)	F (MAX)	L (MIN)
0	0.060	0.096	0.060	0.054	0.050	0.025	0.007	0.50
1	0.073	0.118	0.073	0.066	1/16	0.031	0.007	0.62
2	0.086	0.140	0.086	0.077	5/64	0.038	0.008	0.62
3	0.099	0.161	0.099	0.089	5/64	0.044	0.008	0.62
4	0.112	0.183	0.112	0.101	3/32	0.051	0.009	0.75
5	0.125	0.205	0.125	0.112	3/32	0.057	0.010	0.75
6	0.138	0.226	0.138	0.124	7/64	0.064	0.010	0.75
8	0.164	0.270	0.164	0.148	9/64	0.077	0.012	0.88
10	0.190	0.312	0.190	0.171	5/32	0.090	0.014	0.88
1/4	0.250	0.375	0.250	0.225	3/16	0.120	0.014	1.00
5/16	0.312	0.469	0.312	0.281	1/4	0.151	0.017	1.12
3/8	0.375	0.562	0.375	0.337	5/16	0.182	0.020	1.25
7/16	0.437	0.656	0.438	0.394	3/8	0.213	0.023	1.38
1/2	0.500	0.750	0.500	0.450	3/8	0.245	0.026	1.50
5/8	0.625	0.938	0.625	0.562	1/2	0.307	0.032	1.75
3/4	0.750	1.125	0.750	0.675	5/8	0.370	0.039	2.00
7/8	0.875	1.312	0.875	0.787	3/4	0.432	0.044	2.25
1	1.000	1.500	1.000	0.900	3/4	0.495	0.050	2.50

Note:
1. 1960 series SHCS are typically made from a high grade alloy steel, hardened to a range of 37-45 RC.
2. "F" above is a fillet extension beyond "D".
3. Consult supplier to determine available lengths "L" for each screw size. Typical length increments are as follows:
   1/16" increments - lengths 1/8" thru 1/4"
   1/8" increments - lengths 1/4" thru 1"
   1/4" increments - lengths 1" thru 3.5"
   1/2" increments - lengths 3.5" thru 7"
   1" increments - lengths 7" thru 10"

Shrinkage Values For Miscellaneous Resins

MATERIAL	SHRINK (in/in)
ABS....high impact	0.005-0.007
ABS....heat resistant	0.004-0.005
ABS....medium impact	0.005
ACETAL	0.020-0.035
ACRYLIC....general purpose	0.002-0.009
ACRYLIC....heat resistant	0.003-0.010
ACRYLIC....high impact	0.004-0.008
ETHYLENE VINYL ACETATE	0.010-0.030
IONOMER	0.003-0.020
NYLON....6/6	0.010-0.025
NYLON....6	0.007-0.015
NYLON....6/10	0.010-0.025
NYLON....11	0.010-0.025
NYLON....12	0.008-0.020
NYLON....glass filled	0.005-0.010
POLYBUTYLENE	0.020 (molded)
POLYBUTYLENE	0.040 (aged)
POLYCARBONATE	0.005-0.007
PET (amorphous, petg, pctg)	0.003-0.005
POLYESTER .025-.050 thick	0.006-0.012
POLYESTER .050-.100 thick	0.012-0.017
POLYESTER .100-.180 thick	0.016-0.022
POLYESTER PBT	0.010-0.020
POLYESTER PET 30% GF	0.001-0.002
POLYESTER PBT 30% GF	0.003-0.005
POLYETHERIMIDE	0.005-0.007
POLYETHYLENE....low dens	0.015-0.035
POLYETHYLENE....high dens	0.015-0.030
PPO/STYRENE CO (NORYL)	0.005-0.007
POLYPROPYLENE	0.010-0.030
POLYSTYRENE....G.P.	0.002-0.008
POLYSTYRENE..heat resistant	0.002-0.008
POLYSTYRENE.impact mod	0.003-0.006
POLYSULFONE	0.008
POLYURETHANE	0.010-0.020
PVC....rigid	0.002-0.004
PVC....semi-rigid	0.005-0.025
PVC....flexible	0.015-0.030
SAN	0.002-0.006

Notes:
1. Shrink rates can vary with process conditions, mold design, etc
2. Shrink may be less in direction perpendicular to flow due to differences in orientation (divide by 1.6 for transverse direction; more pronounced with crystalline resins: i.e. PP). Does not apply for glass filled resins.
3. When possible, prototype the job to better determine shrink for a given mold and product design.
4. Refer also to manufacturer's guidelines, if available.
5. Shrink may be affected by additives such as color or fillers.

Calculating Shrinkage & Cavity Sizing

$$\text{Shrinkage} = \frac{(\text{cavity dim} - \text{part length})}{\text{cavity dimension}}$$

$$\text{Cavity Dim} = \frac{\text{finished part length}}{(1 - \text{shrink rate})}$$

Note the following example:

If a part length is designed at 10 inches and the shrink rate is 0.030 in/in; the resulting cavity dimension would be 10.309 instead of 10.300 as one would get by multiplying 10 X 1.030; this is a common mistake made by molders and mold designers.

If a mold is run at very hot temperatures, the mold steel thermal expansion should also be taken into consideration.

Increasing packing pressure can reduce shrinkage; minimizing pressure drop through the mold can also reduce shrinkage (i.e. larger gates, larger runners, etc.) Warmer molds and warmer melt will also reduce pressure drop, but may or may not increase shrinkage: this is because the higher temperatures cause increased shrinkage; usually the higher effective pressure will override the increased shrinkage effects from higher temperature. It is usually more effective to keep temperatures cooler and fill fast to keep pressure drop to a minimum; this also serves to reduce viscosity (discussed in section on relative viscosity).

Common Abbreviations For Misc Resins

RESIN GENERIC NAME	COMMON ABBR
ACRYLONITRILE-BUTADIENE-STYRENE	ABS
CELLULOSE ACETATE	CA
CELLULOSE ACETATE BUTYRATE	CAB
CELLULOSE ACETATE PROPIONATE	CAP
CHLORINATED POLYETHYLENE	CPE
CHLORINATED POLYVINYL CHLORIDE	CPVC
ETHYLENE-PROPYLENE DIENE RUBBER	EPDM
EXPANDABLE POLYSTYRENE	EPS
ETHYLENE VINYL ACETATE	EVA
FIBER REINFORCED PLASTIC	FRP
HIGH DENSITY POLYETHYLENE	HDPE
HIGH IMPACT POLYSTYRENE	HIPS
LOW DENSITY POLYETYLENE	LDPE
LINEAR LOW DENSITY POLYETYLENE	LLDPE
MALEIC ANHYDRIDE	MA
MEDIUM DENSITY POLYETHYLENE	MDPE
POLYAMIDE (NYLON)	PA
POLYACRYLONITRILE	PAN
POLYBUTYLENE	PB
POLYBUTYLENE TEREPHTHALATE	PBT
POLYCARBONATE	PC
POLYETHYLENE	PE
POLYETHERETHERKETONE	PEEK
POLYETHERIMIDE	PEI
POLYETHERKETONE	PEK
POLYETHYLENE TEREPHTHALATE	PET
POLYETHYLENE TEREPHTHAL GLYCOL	PETG
POLYIMIDE	PI
POLYMETHYL METHACRYLATE	PMMA
POLYOXYMETHYLENE	POM
POLYPROPYLENE	PP
CHLORINATED POLYPROPYLENE	PPC
POLYPHENYLENE OXIDE	PPO
POLYPHENYLENE SULFIDE	PPS
POLYPHENYLENE SULPHONE	PPSU
POLYSTYRENE	PS
POLYSULFONE	PSU
POLYURETHANE	PU
POLYURETHANE	PUR
POLYVINYL CHLORIDE	PVC
REINFOR POLYBUTYLENE TEREPHTHAL	RPBT
REINFOR POLYETHYLENE TEREPHTHAL	RPET
REACTION INJECTION MOLDING	RIM
STYRENE ACRYLONITRILE	SAN
STYRENE BUTADIENE RUBBER	SBR
STYRENE MALEIC ANHYDRIDE	SMA
SHEET MOLDING COMPOUND	SMC
THERMOPLASTIC ELASTOMER	TPE

A Method For Estimating The Cooling Cycle

$$Q = \frac{-(t^2)}{2\pi\alpha} \times LN\left[\frac{\pi(T_x-T_m)}{4(T_c-T_m)}\right]$$

See list below for explanations of terms; see table farther below for values.
Q = cooling time (sec) ... t = part thickness (inches)
α = thermal diffusivity (in²/sec) = K ÷ ρ Cp
K = thermal conductivity (Btu/hr ft °F) ... ρ = density (lb/ft³)
Cp = specific heat (Btu/lb°F)
Tx = heat deflection temperature (use table less $\cong$ 30°)
Tm = mold temperature (°F) ... Tc = cylinder or melt temperature (°F)
LN = natural logarithm

NOTE: Cooling time using this formula becomes somewhat conservative or long for thicker part walls. This is because actual moldings of thick walled parts typically result in opening the mold prior to full wall cooling (a hotter or slightly molten center may exist on thick walled parts). Post mold part handling is critical in such scenarios.

THERMAL PROPERTIES FOR SELECTED RESINS

RESIN (abbr)	Thermal conductivity K[1] (Btu/hr ft °F)	Cp[1] (Btu/lb°F)	Density[1] (gr/cc)	lb/ft³	Thermal diffusivity[1] (in²/sec)	Deflection Temp[1] @ 66 psi
ABS	0.108 - 0.192	0.490	1.060	66.144	0.000185	203
CA, CAP	0.100 - 0.192	0.410	1.260	78.624	0.000181	192
CAB	0.100 - 0.192	0.390	1.200	74.88	0.000200	201
HIPS	0.024 - 0.073	0.501	1.050	65.52	0.000059	185
IONOM	0.142 - 0.142	0.645	0.950	59.28	0.000148	125
LDPE	0.192 - 0.192	0.760	0.920	57.408	0.000176	113
MDPE	0.192 - 0.242	0.765	0.935	58.344	0.000194	155
HDPE	0.267 - 0.300	0.870	0.960	59.904	0.000217	186
PA6 GF	0.173 - 0.173	0.669	1.380	86.112	0.000120	460
PA 6, 6/6	0.142 - 0.142	0.731	1.140	71.136	0.000109	356
PC	0.108 - 0.108	0.438	1.200	74.88	0.000132	280
PET	0.144 - 0.144	0.502	1.330	82.992	0.000138	153
PET (C)	0.144 - 0.144	0.548	1.360	84.864	0.000124	252
PMMA	0.092 - 0.142	0.454	1.190	74.256	0.000138	215
POM	0.133 - 0.133	0.715	1.420	88.608	0.000084	336
PP	0.066 - 0.079	0.667	0.900	56.16	0.000077	204
PP co	0.048 - 0.100	0.640	0.900	56.16	0.000083	184
PPO/PS	0.125 - 0.125	0.519	1.070	66.768	0.000144	234
GF PPS	0.167 - 0.259	0.497	1.650	102.960	0.000166	500
PS g.p.	0.058 - 0.080	0.480	1.060	66.144	0.000087	180
PSU	0.150 - 0.150	0.520	1.240	77.376	0.000149	345
PVC	0.083 - 0.083	0.383	1.290	80.496	0.000107	156
PVC rig	0.092 - 0.092	0.340	1.400	87.36	0.000123	174
SAN	0.070 - 0.070	0.471	1.080	67.392	0.000088	225

[1] Increased accuracy will result if actual data for your specific resin is obtained from your resin supplier due to variation between different blends from different suppliers. Thermal diffusivity is calculated from density (lb/ft³), specific heat (Btu/lb°F) & thermal conductivity (Btu/hr ft °F). Note: These aforementioned units must be used together for proper unit cancellation to get correct answer. You must convert metric units or use only metric units for proper unit cancellation.

Molding Area Diagram (MAD)

The diagram below is a graphical depiction of how the mold performs for a given product at various levels of melt temperature and packing pressure. These two process variables are very important, but other process parameters are also important. The process generating this plot should be held constant with the exception of melt temperature and packing pressure. The plot could be shifted with changes in mold temperature, injection fill rate and/or other process parameters. Typically the fastest possible fill rate for machine selected is selected and used.

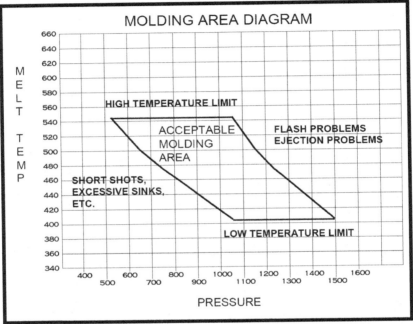

The size of the MAD indicates how forgiving the mold is to process variation (melt temperature and pressure). Some molds with complex and hard to fill parts have a narrow molding window. There may be a fine balance between getting the part filled and packed properly versus overpacking, whereby the part sticks in the mold or cracks on ejection. The mold might be made more forgiving by improving design.

New Mold Debug Checklist

EASE OF MOLD LIFTING AND SETTING:
1. Will mold fit the end use molding press?
2. Mold should have eyebolt holes & safety strap.
3. Mold should hang level to align with locating ring & KOs.
4. Mold should offer protection to external wiring, switches, fittings, etc on bottom of mold.
5. Mold should have clamping slots positioned such that sufficient clamps can be used to secure mold to platens.
6. All water fittings should be positioned so that not in interference with mold clamps, machine doors, tie bars, etc.
7. Mold should have sufficient eyebolts so that mold rotation can be achieved with a second eyebolt and hoist.
8. Sprue radius should be compatible, sprue recess ID should be compatible with machine's nozzle/heater band OD.
9. Mold should have total weight listed on clamping plate.
10. Socket head cap screws retaining clamping plate should be slightly below flush to prevent platen damage.
11. Electrical receptacles should be accessible & away from water fittings.

BASIC MECHANICAL FUNCTIONALITY:
12. Water lines should be free flowing; check with flow meter/indicator or direct return into bucket to observe flow (do not try to measure/quantify with bucket - not accurate due to lack of back pressure).
13. Ejector plate should return freely and completely.
14. Slides should move freely, but be retained during mold open in the pulled position.
15. Leader pins should not exhibit galling.
16. All moving slide surfaces and slide locks should be free from galling.

ELECTRICAL FUNCTIONALITY:
17. Know the heater wattage; calculate the full load amperage and compare to actual (or at least compare to other like heaters).
18. Each T/C should indicate a proper response for it's heater.
19. Heater cables should be grounded at mold & controller ends.

LOCATION ITEMS:
20. Ejector pins should be flush to 0.001" below flush.
21. Examine parts looking for long ejector pins.
22. Examine parts/runner looking for parting line mismatch.
23. Examine parts looking for slide mislocation.
24. Check part for proper wall thickness with pointed micrometers.
25. Examine wall thickness variation relative to fill problems.
26. Intentional drags should be adjacent to ejector pins and be shallow @ 0.003", but sharp (can then reduce sharpness if too much drag).
27. Alignment should be located on horizontal and vertical centerlines.

ESTABLISH A BASIC PROCESS:
28. Record all pertinent process & setup data.
29. Determine gate seal time.

DETERMINE BALANCE OF FILL:
30. Establish basic process.
31. Set hold/pack pressure to zero.
32. Adjust inject time or transfer position to achieve only one full part (or as close as possible).
33. Collect 3 shots; separate parts by cavity and determine avg part weight for each cavity.
34. Max imbalance should be 15% or less.

MISC OTHER CHECKS:
35. Have all sharp corners removed from mold base exterior?
36. Do parts fall free and clear of mold without knicks?
37. Have pry bar slots been installed in main parting lines?
38. Review gate vestige relative to requirements.
39. Review for proper mold surface finish from both cosmetic and functionality standpoint.
40. Lightly touch each core and cavity just after repeated cycling looking for hot spots (Do not touch hot molds; observe standard safety practices before reaching into press).
41. Stop mold prior to any ejection, look for raised surfaces where part may be trying to stick in non-moving half.
42. Look at parts for parting line drags.
43. Perform dimensional checks after shrinkage (48 hrs).

Sample Mold Information Sheet

Mold Information Sheet

MOLD DESCRIPTION					
NUMBER OF CAVS.		MOLD DRWG NUMBER			
PART DESCRIPTION					
RUNNER TYPE					
MOLD SIZE	HORIZ	VERT		SHUT HT	WT

SPRUE RADIUS		SPRUE ORIFICE	
RECESS I.D.		RECESS DEPTH	
EJECTION TYPE			
K.O. PATTERNS			
K.O. THREAD			
GATE SIZE/TYPE			
CAVITY MATERIAL		HARDNESS	Rc
CORE MATERIAL		HARDNESS	Rc
POWER CONSUMPTION (watts)	MANIFOLD		
	GATE DROPS		

SEE CAVITY LOCATIONS BELOW
TOP/EJECTOR HALF   00

```
              |  13    9  |   |   5    1  |
OPERATOR SIDE |  14   10  |   |   6    2  |
              |  Manifold 2   |   Manifold 1
              |  15   11  |   |   7    3  |
              |  16   12  |   |   8    4  |
```

Adopt a standard cavity numbering sequence such as top to bottom starting at offset corner (back side, top in this example).

Since the HR system is located in A-half, the layouts typically start in top left when looking at parting line; thus, this layout. This view is B half where you grab the parts!

Page 30

Sample Mold Process Sheet

Molding Conditions Record

DATE/TIME: _____ PART: _____ PART WT (GRAMS): _____
PART DESCRIPTION: _____ MOLD NO: _____ SHOT WT (GRAMS): _____
REV # OR ECN (IF APPLICABLE): _____ NO. CAVITIES: _____
MACHINE ID: _____ TONS: _____ SHOT SIZE (OZ): _____
MECH ADV: _____ SCREW TYPE: _____ NOZZLE ORIFICE ID: _____
RUN BY: _____
CORE SEQUENCE: _____ CIRCLE EJECT TYPE: PUSH ONLY   PUSH/PULL   PUSH/SPRING RETURN   OTHER

	RUN NO. OR ACTIVITY DESCRIPTION >>>>>						
MATERIAL	MATERIAL						
	LOT NO.						
	COLOR CONC						
	COLOR BLEND RATIO						
	OTHER ADDITIVES						
DRYING	DEW POINT	(°F) (°C)					
	TEMPERATURE	(°F) (°C)					
	TIME	(HOURS)					
	HOPPER SIZE						
	CALCULATED NORMAL RESIDENCE TIME						
TEMPERATURES	FEED ZONE	(°F) (°C)					
	CENTER ZONE	(°F) (°C)					
	FRONT ZONE	(°F) (°C)					
	NOZZLE	(°F) (°C)					
	HOT MANIFOLD	(°F) (°C)					
	ACTUAL MELT:	(°F) (°C)					
	MOLD - FIXED	(°F) (°C)					
	MOLD - MOVEABLE	(°F) (°C)					
	MOLD - SLIDES	(°F) (°C)					
PRESSURES	FILL VELOCITY	(%)					
	CLAMP	(%) (TONS)					
	FILLING PRESSURE	P1					
	PACKING PRESSURE	P2					
	HOLD PRESSURE	P3					
	BACK PRESSURE						
CYCLE TIMES	INJ. TOTAL SCREW FORWARD	(SEC)					
	FILL TIME	(SEC)					
	PACK TIME	(SEC)					
	HOLD TIME	(SEC)					
	COOLING	(SEC)					
	PLASTICIZING	(SEC)					
	MOLD OPEN TIMER	(SEC)					
	ACTUAL TOTAL OPEN	(SEC)					
	OVERALL	(SEC)					
	RESIDENCE TIME - BARREL (MIN.)	CALC					
MISCELLANEOUS	TRANSFER METHOD						
	TRANSFER POSITION	(INCHES) (MM)					
	TRANSFER WEIGHT	(%)	CALC				
	DECOMPRESSION	(INCHES) (MM)					
	SHOT SIZE	(INCHES) (MM)					
	CUSHION	(INCHES) (MM)					
	END OF STROKE	(INCHES) (MM)					
	SCREW SPEED	(RPM) (%)					
COMMENTS	A -						
	B -						
	C -						
	D -						
	E -						

Melt Index Test

Melt index measures flowability and is affected by molecular weight and/or weight distribution. The melt index number value is simply the weight in grams of the amount of plastic extruded in a 10 minute test period. It should be noted there are numerous test conditions (different weights and temperatures); thus, caution is needed when comparing melt index numbers if the test condition is unknown. See table below for a listing of various test conditions. When testing hygroscopic resins, the preparation of the sample is critical in terms of accomplishing good repeatable drying. Note: Test done at very low shear rates, unlike normal injection molding.

Standard Test Conditions			
Condition	Temp / Load (°C / Kg)	(kPa)	Pressure (psi)
A	125 / 0.325	44.8	6.50
B	125 / 2.16	298.2	43.25
C	150 / 2.16	298.2	43.25
D	190 / 0.325	44.8	6.50
E	190 / 2.16	298.2	43.25
F	190 / 21.6	2982.2	432.50
G	200 / 5.0	689.5	100.00
H	230 / 1.2	165.4	24.00
I	230 / 3.8	524.0	76.00
J	265 / 12.5	1723.7	250.00
K	275 / 0.325	44.8	6.50
L	230 / 2.16	298.2	43.25
M	190 / 1.05	144.7	21.00
N	190 / 10.0	1379.0	200.00
O	300 / 1.2	165.4	24.00
P	190 / 5.0	689.5	100.0
Q	235 / 1.0	138.2	20.05
R	235 / 2.16	298.2	43.25
S	235 / 5.0	689.5	100.00
T	250 / 2.16	298.2	43.25
U	310 / 12.5	1723.7	250.00
V	210 / 2.16	298.2	43.25
W	285 / 2.16	298.2	43.25
X	315 / 5.0	689.5	100.00

Resin	Test Conditions
Acetals	E, M
Acrylics	H, I
ABS	G, I
Cellulosics	D, E, F, V
Nylon	K, Q, R, S
PE	A, B, D, E, F, N, U
PC	O
PP	L
PS	G, H, I, P
PET	T, V, W
PVC	C
PPS	X

Physical Properties Of Resins: ASTM Testing

The following is a very brief description of some important ASTM tests and the purpose of the data.

ASTM D955 SHRINKAGE:
Specimens molded are very simple shapes. The part dimensions are compared to the mold dimensions in order to compute shrinkage. Data is a rough guideline only, because the test bar complexity and process do not closely compare to actual molding/manufacturing. When close tolerance designs exist, it is best to prototype or gather data from similar jobs.

ASTM D1238 MELT INDEX:
Resin is heated at a specified temperature; a standard weight (usually 2160 grams) is on top of a piston which extrudes the melt at very low shear rates through a 2.1 mm orifice. The grams per 10 minutes is determined and termed the melt index. This test indicates the flowability of a given resin, but does not indicate total relative flowability - due to shear rate at which test is done. It should also be noted that many "conditions" exist for performing the test which require different loadings and different temperatures.

ASTM D638 TENSILE PROPERTIES:
Molded, dog bone shaped, specimens are held at each end and pulled apart. The test indicates strength of the material as well as how much the material will stretch before breaking. The elastic modulus, or tensile modulus, is the ratio of stress to strain below the proportional limit (below the elastic limit which is the greatest stress without permanent strain). The stress is plotted versus strain (elongation). A pull rate of 0.2 inches/minute is often used, but faster rates are performed. The rate of pull effects the results; thus, different speeds are done and compared. The area under the plotted stress-strain curve indicates overall toughness.

ASTM D790 FLEXURAL PROPERTIES:
Flexural strength indicates the maximum stress that a material sustains at the moment of break. The test bar specimen (typically a 1/8 X 1/2 X 5 in.) is supported near each end. A force is applied to the middle. Many plastics do not break; thus, the flexural strength is the stress level at 5% strain. The dynamics of this test evaluate both the compressive and tensile strengths of the test bar - compressive on the top or concave surface and tensile on the bottom or convex surface. The flexural modulus is the ratio, within the elastic limit, of stress to corresponding strain.

ASTM D526 IZOD IMPACT TEST:
The Izod impact test indicates the energy required to break specimens under standard test conditions. Test bars are molded with a notch and typically sized at 1/8 X 1/2 X 2 inches. The test bar is placed vertically in a clamp so that it is cantilevered upward. A pendulum swings into the test bar; the impact strength is calculated from the height of the pendulum follow through as it continues it's swing. The test data is best used to compare various grades of resin within the same resin family. Indication of absolute toughness is misleading because some resins are notch sensitive.

ASTM D648 DEFLECTION TEMPERATURE:
Test bars are supported on each end (4 inches between supports) and a load of either 66 or 264 psi is placed on the center. The temperature is raised at a rate of 2 degrees C per minute. When the bar has deflected 0.010 inches, that temperature is recorded as the deflection temperature.

Density Values - Selected Resins

MATERIAL	70° F DENSITY (gr/cc)	MELT DENSITY (gr/cc)
ABS (INJECTION GRADE)	1.05	0.97
ABS....30% GF	1.28	
ACETAL	1.42	1.17
ACETAL....20% GF	1.55	
ACRYLIC....general purpose	1.16	1.04
CELLULOSE ACETATE	1.26	1.13
CELLULOSE BUTYRATE	1.20	1.07
CELLULOSE PROPIONATE	1.22	1.10
ETHYLENE VINYL ACETATE	0.95	
IONOMER	0.95	0.73
NYLON....6/6	1.14	0.97
NYLON....6	1.13	0.97
NYLON....6/10	1.08	0.97
NYLON....6/12	1.07	0.97
NYLON....11	1.04	0.97
NYLON....12	1.02	0.97
NYLON....30% GF	1.36	
POLYBUTYLENE	0.92	
POLYCARB/ESTER ALLOY	1.20	
POLYCARBONATE	1.20	1.02
POLYCARBONATE....30% GF	1.42	
CO-POLYESTER PETG	1.27	1.12
CO-POLYESTER PCTG	1.24	1.08
POLYESTER PBT	1.34	1.11
POLYEST PBT/PET 30% GF	1.58	
POLYESTER PET (bottle)	1.40	1.20
POLYETHERIMIDE	1.27	
POLYETHYLENE...low density	0.92	0.76
POLYETHYLENE...med density	0.92 - 0.94	0.74
POLYETHYLENE...high density	0.94 - 0.97	0.72
POLYETHYL...30% GF HDPE	1.18	
PPO/STYRENE (NORYL)	1.07	0.90
POLYPHENYLENE SULFIDE	1.35-1.80	
POLYPHEN SULFIDE.40% GF	1.65	
POLYPROPYLENE	0.90	0.70
POLYPROPYLENE....30% GF	1.13	
POLYSTYRENE....G.P.	1.05	0.97
POLYSTYRENE....impact mod	1.04	0.96
POLYSULFONE	1.24	1.16
POLYURETHANE	1.20	1.13
PVC - Rigid	1.39	1.30
PVC - Pipe	1.44	1.32
PVC - Flexible	1.29	1.20
SAN	1.08	1.00
T/P RUBBER.......(Santoprene)	0.97	0.93
T/P ELASTOMER..(Urethane)	0.83	0.82

Thermoplastic Identification

BURN AND SNIFF TESTS

Drop a sample in water — floats [PE, PP] — sinks [all others]

Resin	Color of Flame	Odor	Burn Speed	Other
PE	blue, yellow tip	paraffin	fast	melts & drips
PP	blue, yellow tip	diesel fumes	slow	
ABS	blue, yellow edge	acrid	slow	black smoke w/ soot
POM	blue	formaldehyde	slow	no smoke
CA	yellow w/sparks	vinegar	slow	black smoke w/ soot
CAB	yellow w/blue tip	rancid butter	slow	smoke w/ soot
CAP	yellow	burnt sugar	fast	some black smoke
PS	yellow	gas or marigold	fast	dense smoke & soot
PET	yellow	burning rubber	fast	black smoke w/soot
PU	yellow	faint apple	fast	some black smoke
self extinguishing and drips				
PA	blue, yellow tip	burnt hair	slow	froths
PSU	orange	sulphur	fast	black smoke w/soot
PC	orange or yellow	phenol	slow	black smoke w/soot
self extinguishing, but no drips				
PPO	yellow orange	phenol	slow	difficult to ignite
PVC	yellow w/green edge	hydrochloric	slow	white smoke

NOTE: Burn & sniff tests are old school and not recommended for health & safety reasons, use other lab tests (such as IR Spectroscopy listed below) to identify unknown resins.

IR Spectroscopy

In this testing, the sample is exposed to infrared light. The absorption pattern of this light is recorded; the resulting absorption pattern graph yields a "fingerprint" of the material. IR is used to identify different plastics. The following sample is polycarbonate; the actual scan is compared to a reference scan to determine the resin's identity. In this case the actual is nearly identical to the reference indicating the resin in question is polycarbonate.

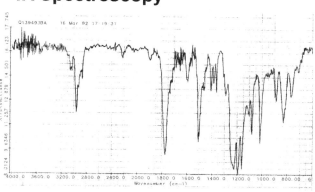

REFERENCE SCAN

18162-5
Polycarbonate resin

IR III, 1589C

2968.9 1289.6 1015.0
1774.4 1192.9 830.7
1505.5 1080.8 556.7

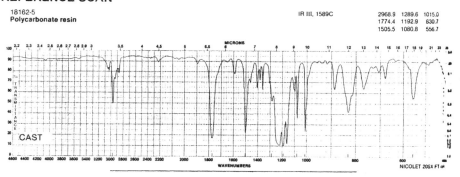

Page 35

Melt Temperature Control and Measurement

Several points should be discussed regarding melt temperature:
1. Develop a habit of checking heater bands and thermocouples for proper operation. T/Cs should reach bottom of "well" or hole and be preloaded (spring loaded) against barrel. Keep T/C and well clean from rust, scale & residue. Use T/C wire and plugs for extensions.
2. Rotate screw speed only fast enough to get back just prior to mold open (unless nozzle valve is present, then screw rotation may continue during mold opening if hydraulics will permit); this will improve the melt thermal homogeneity.
3. Use some back pressure - even as low as 20-30 psi, but 50 psi works well to enhance consistency.
4. The barrel temperature profile is often thought to be optimum when the extruder pressure is approaching 50-75% of maximum.
5. The front zone should neither heat nor cool the melt, but be set equal to the actual measured melt temperature.
6. Check the actual melt temperature by preheating a needle type thermocouple probe to approximately 30 degrees above the estimated actual. Collect melt shot into a collection cup/container. Stir needle probe in melt slowly until temperature stabilizes. Note: Select the smallest probe possible to minimize the mass required to heat, but large enough to endure the usage.
7. Crystalline resins typically require higher rear zone barrel setpoints (a descending profile may yield faster screw recovery).

Resin Process Temperatures

PROCESS MATERIAL	MELT TEMP (°F)	MOLD TEMP (°F)
ABS....med, high impact	440-510	90-180
ABS....high heat grade	510-540	140-180
ACETAL	380-420	140-220
ACRYLIC	420-485	120-180
CA, CAB	360-440	80-130
ETHYLENE VINYL ACETATE	350-400	50-100
IONOMER	420-460	50-100
NYLON....6/6	500-560	120-180
NYLON....6	470-530	100-180
NYLON....6/10	460-510	100-180
NYLON....11	420-480	80-140
NYLON....12	400-450	80-140
POLYCARB....lo/med viscosity	540-570	160-200
POLYCARB....high viscosity	590-640	180-240
POLYESTER - PBT	460-490	100-140
POLYESTER - PET (Bottle)	520-560	60-120
POLYESTER - PETG	480-520	70-120
POLYESTER - PCTG	520-560	70-120
POLYESTER - PCT (GF)	565-590	200-250
POLYESTER - PCTA (GF)	560-590	300-320
POLYETHERIMIDE	680-720	220-300
POLYETHYLENE....low density	340-440	50-100
POLYETHYLENE....med density	390-490	50-120
POLYETHYLENE....high density	420-540	50-150
PPO/STYRENE COPOLYMER	480-580	150-220
POLYPHENYLENE SULFIDE	590-670	190-230
POLYPROPYLENE	420-520	60-150
POLYSTYRENE....G.P.	380-460	50-120
POLYSTYRENE...impact modified	400-480	50-120
POLYSULFONE	650-750	200-320
POLYURETHANE	390-440	70-120
PVC....Flex - Rigid	320-420	50-120
SAN	400-500	120-180
T/P ELASTOMER	340-440	70-120

Notes:
1. Flame retardant grades can be 50° less.
2. Follow manufacturer's guidelines if available.
3. Glass or mineral fillers may require higher heats.
4. Amorphous polyesters may stick to hot steel > 150° F.
5. Long residence time may require lower barrel setpoints.

Calculate Heat Load and Cooling Required

$$\frac{\text{Heat Load (resin thruput} \times \text{spec heat (resin)} \times \Delta T)}{\text{spec heat (coolant)} \times \text{temp rise of coolant}} = \text{lbs / hr of coolant required}$$

$$\frac{\frac{\text{lbs}}{\text{hr}} \text{ of coolant required}}{\text{coolant} \frac{\text{lbs}}{\text{gal}} \times 60 \frac{\text{min}}{\text{hr}}} = \text{GPM of coolant required}$$

EXAMPLE CONDITIONS (see calculation below)
shot wt = 221.5 grams (0.487 Lbs)
cycle time = 15 sec
resin = PP copolymer
spec heat PP = 0.64 Btu/(Lb °F)
spec heat water = 1.003 Btu/(Lb °F) ... (@ 55°F)
ejected part temp = 120 °F
melt temp = 440 °F
ΔT = 320 °F
temp rise in water = 5 °F (this is max allowable temp rise; 2-3° F is suggested)
1 gal water = 8.33 lbs ... 1 Ton chiller = 12,000 Btu/Hr

$$\frac{0.487 \text{ lbs}}{15 \text{ sec}} \times \frac{3600 \text{ sec}}{\text{hr}} = 116.88 \frac{\text{lbs}}{\text{hr}}$$

$$116.88 \frac{\text{lbs}}{\text{hr}} \times \frac{0.64 \text{ Btu}}{\text{lb °F}} \times 320 \text{ °F } (\Delta T) = 23,937 \frac{\text{Btu}}{\text{hr}}$$

$$\frac{23,937 \frac{\text{Btu}}{\text{hr}}}{1.003 \frac{\text{Btu}}{\text{lb °F}} \times 5 \text{ °F}} = 4,773 \frac{\text{lbs}}{\text{hr}} \text{ of coolant}$$

$$\frac{4,773 \frac{\text{lbs}}{\text{hr}}}{8.33 \frac{\text{lbs}}{\text{gal}} \times 60 \frac{\text{min}}{\text{hr}}} = 9.54 \text{ GPM of coolant reqd}$$

$$\frac{23,937}{12,000} = 2 \text{ tons of chiller capacity required}$$

NOTE: If all lines checked collectively together, their is risk that the GPM may not be present where heat load is present (specifically at core/cavity cooling circuits). This relates to disadvantage in connecting too many INS & OUTS if sufficient water is not available to keep them all flowing at high rate. Flow indicator/regulators work well at indicating free flowing channels where little work is done; these can be throttled down to increase flow thru channels with higher heat loads. A high percentage of available water will take the path of least resistance thru free flowing channels (e.g. clamp plates and manifold plates).

Calculating Reynolds Number

It is important for the coolant flowing thru a mold to be in turbulent flow so that the coolant nearest the O.D. of the water line doesn't do all the work; turbulent flow allows more of the water to become exposed to the core/cavity coolant channels. Reynolds number is the number which indicates if and when the flow is turbulent. A Reynolds number (Nr) of 10,000 is desired for optimum cooling. A quick rule of thumb to verify proper Reynolds number (and turbulent flow) is having flow (GPM) equal to 3.5 times the coolant channel ID (round or equivalent round). Note: Equiv hyd diam = 4A/P (a = area; p = perimeter).

$$Nr = \frac{7740VD}{n} \text{ or } \frac{3160}{Dn}$$

V = fluid velocity in ft/sec
D = diameter of passage in inches
n = kinematic viscosity in centistokes
Q = coolant flow rate in gpm

TABLE of KINEMATIC VISCOSITY FOR WATER	
°F	VISCOSITY CENTISTOKES
32	1.79
40	1.54
50	1.31
60	1.12
70	0.98
80	0.86
90	0.76
100	0.69
120	0.56
140	0.47
160	0.40
180	0.35
200	0.31
212	0.28

NOTES:
1. The chart (right) is for water; an ethylene/glycol mix changes the viscosity and may result in laminar flow unless very high flow rates are achieved. Ethylene/glycol has lower thermal conductivity and a lower specific heat (0.575 vs 1); thus, carrying less heat out of mold.
2. Attempting to connect all water lines as separate INS and OUTS will normally result in lower flow rates thru each individual circuit (versus some use of loops or bypass lines). These reduced flow rates may result in reduced heat transfer. Heat transfer coefficients calculated using Reynolds, Prandtl and Nusselt numbers continue to increase as Reynolds number increases (but, heat transfer assumes the heat conducts to the cooling channel wall ID as fast as the coolant can remove it). Therefore, it is suggested to use loops as needed to make mold setup easier and to keep flow rates elevated, but temperature rise in the circuit should not exceed 5° F (1–3° F is normal and suggested).
3. The Nr is valid only for the location for which it is calculated ... meaning a specific core/cav location. DO NOT calculate a Nr for a given entrance water line size if the size changes elsewhere in the mold where the work is done.
4. A water line in parallel should have its actual flow rate recalculated accordingly (if the measured flow is prior to the parallel branching).

API Schedule 40 & 80 Pipe Data

WATER FLOW IN PIPES
Water flow in a pipe can be approximated by the following equation, but a velocity of 5-8 ft/sec is often used instead for planning purposes (e.g. take velocity such as 7 ft/sec X CSA to get ft^3/sec for conversion to GPM).

$$V = C \times \sqrt{\frac{hD}{L + 54D}}$$

V= mean velocity (ft/sec)
C= coefficient (see table right)
D= diameter of pipe (ft)
h= total head (ft) ... (1 psi = 2.309 ft head)
L= total length of pipe line (ft)

(APPLIES TO STEEL AND PVC)

PIPE SIZE (IN)	SCHD NO.	ID (IN)	CSA (FT^2)	OD (IN)	WALL (IN)	COEFF
1/8	40	0.269	0.0004	0.405	0.068	
1/4	40	0.364	0.0007	0.540	0.088	
3/8	40	0.493	0.0013	0.675	0.091	
1/2	40	0.622	0.0021	0.840	0.109	
3/4	40	0.824	0.0037	1.050	0.113	
1	40	1.049	0.0060	1.315	0.133	20.00
1	80	0.957	0.0050	1.315	0.179	19.5
1-1/4	40	1.380	0.0104	1.660	0.140	24.05
1-1/4	80	1.278	0.0089	1.660	0.191	23.46
1-1/2	40	1.610	0.0141	1.900	0.145	25.39
1-1/2	80	1.500	0.0123	1.900	0.200	24.75
2	40	2.067	0.0233	2.375	0.154	28.06
2	80	1.939	0.0205	2.375	0.218	27.31
2-1/2	40	2.469	0.0332	2.875	0.203	30.23
3	40	3.068	0.0513	3.500	0.216	32.23
3	80	2.900	0.0459	3.500	0.300	31.67
4	40	4.026	0.0884	4.500	0.237	35.07
4	80	3.826	0.0798	4.500	0.337	34.57
6	40	6.065	0.2006	6.625	0.280	39.16
6	80	5.761	0.1810	6.625	0.432	38.60
8	40	7.981	0.3474	8.625	0.322	43.30
8	80	7.625	0.3171	8.625	0.500	42.71

Once velocity in ft/sec is known: multiply it by CSA in ft^2 to get ft^3/sec; multiply ft^3/sec X 7.48 X 60 for GPM.
Head losses should be added to "L" as follows:
90° Elbow has an L value equal to 1.333D
Valves have L value equal to 6D
Approximate HP REQD = (PSI X GPM)/1714 X 0.70
NOTE: Refer to specific pump curves from supplier to more accurately predict GPM & PSI output from a given pump and impeller.

COOLING TOWER REQUIREMENTS	
Air compressor	0.2 ton/HP
Hot runner system	0.25 ton/KW
Mach hyd heat exchanger	0.1 ton HP (1 HP = 0.746 KW)
Barrel cooling	1 ton/inch screw diam
Feedthroat cooling	0.5 ton/press
Plan tower pump to have 3 GPM/TON tower capacity	

Basic Algebra

<u>Commutative law</u>: numbers can be added or multiplied in any order:
$a+b = b+a$ OR $ab = ba$

<u>Associative law</u>: the sum or product of three or more terms is unaffected by the grouping of the terms:

$a+b+c = a+(b+c) = (a+b)+c$ ... OR ... $abc = a(bc) = (ab)c = (ac)b$

<u>Distributive law</u>: $a(b+c) = ab + ac$

<u>Operations with zero or negative numbers</u>:

$a+(-a) = 0$ ... $-(-a) = a$ ... $a \times 0 = 0$ ... $a(-b) = -ab$ ...
$0/a = 0$, $0 \div a = 0$ (if a not equal to zero) ... $(-a)(-b) = ab$

<u>Powers or Exponents</u>:

$$a^m \times a^n = a^{m+n}$$

$$\frac{a^m}{a^n} = a^{m-n} \qquad (ab)^n = a^n b^n$$

$$a^{-n} = \left[\frac{1}{a}\right]^n = \frac{1}{a^n}$$

$$(a^m)^n = a^{mn} \qquad a^0 = 1;\ 0^n = 0$$

Fractions:

$$\frac{a}{c} \pm \frac{b}{d} = \frac{ad \pm bc}{cd}$$

$$\frac{a}{c} \pm \frac{b}{c} = \frac{a \pm b}{c}$$

$$\frac{a}{c} \pm \frac{a}{d} = \frac{a(d \pm c)}{cd}$$

$$\frac{a}{b} \times \frac{c}{d} = \frac{ac}{bd}$$

$$\frac{\frac{a}{b}}{\frac{c}{d}} = \frac{a}{b} \times \frac{d}{c} = \frac{ad}{bc}$$

Conversion Factors

LENGTH
1 inch = 25.4 mm
1 mm = 0.03937 in
1 foot = 30.48 cm
1 micron = 0.001 mm
1 micron = 0.0000394 in
1 inch = 2.54 cm
1 meter = 39.37 in
1 meter = 100 cm
1 microinch = 0.000001 in
1 microinch = 0.0254 microns
(printer's)
1 pica = 0.166 in
1 point = 0.01384 in

WEIGHT
1 lb = 453.6 gr
1 lb = 16 oz
1 gram = 0.035 oz
1 kg = 1000 gr
1 kg = 2.2046 lb
1 oz = 28.35 gr
1 metric ton = 2204.6 lb
1 metric ton = 1000 kg

ANGLES
1 degree = 0.01745 radian
1 degree = π/180 radian

VOLUME
1 cu in = 16.387 cc
1 cu ft = 1728 cu in
1 qt = 0.946 L
1 gal = 128 oz
1 cc = 1 gr (water)
1 gal = 8.33 lb
1 cu ft = 7.48 gal

AREA
1 sq in = 6.452 cc
1 sq ft = 144 sq in
1 acre = 43560 sq ft
1 sq cm = 0.155 sq in
1 sq ft = 0.111 sq yd
1 sq mm = 0.00155 sq in
1 sq km = 0.3861 sq mi

PRESSURE
1 in Hg = 13.6 in H_2O
1 kg/cm^2 = 14.223 psi
1 bar = 14.5 psi
1 atmos = 14.696 psi
1 MPa = 145 psi

ENERGY
1 BTU = 777.97 ft lb
1 cal = 3.09 ft lb
1 BTU = 252 cal
1 kwh = 3412 BTU
1 H.P. = 746 watts
1 ton (refrig) = 12000 Btu/hr
1 ton (refrig) = 3517 watts

SPECIFIC HEAT & HEAT TRANSFER
1 Cal/sec cm °C = 2903 BTU-in/hr ft^2 °F
1 Cal/sec cm °C = 241.9 BTU/hr ft °F
1 W/(m °K) = 0.0023884 Cal/sec cm °C
1 W/(m °K) = 0.5778 BTU/hr ft °F
1 W/(m °K) = 6.9335 BTU-in/hr ft^2 °F
1 Btu/(Lb °F) = 4.184 KJ/(Kg °K)
1 Btu/(Lb °F) = 4184 J/(Kg °K)
1 Cal/(g °C) = 1 Btu/(Lb °F)

TEMPERATURE CONVERSIONS
°C = (°F-32)/1.8
°F = (°C x 1.8) + 32
°K = (°F+459.67)/1.8

Basic Conversions

Conversion factors are very easy to use. The following examples and comments should make the use of conversion factors easier to accomplish. When converting, we start with anything that we know. If you are assigned the task of packaging 3 pounds of plastic for shipment, but the scales only read in grams; we could start with what we know - 3 lbs ... 3 lbs = How many grams?

At this point, it should be pointed out that the conversion factors seen on previous page can be re-written as (see example right):

$$1 lb = 453.6 \, gr = \frac{453.6 \, gr}{1 lb} = \frac{1 lb}{453.6 \, gr}$$

If what is on top is equivalent to what is on bottom ... anything goes.

$$1 lb = 453.6 \, gr = \frac{453.6 \, gr}{1 lb} = \frac{16 \, oz}{1 lb}$$

Now, back to our problem: if we start with lbs and we want to end up with grams, we use a conversion factor to convert.

We will multiply 3 lbs by a conversion factor: $\frac{453.6 \, gr}{1 lb}$

We can rewrite the above as: $\frac{3 \, lb}{1} \times \frac{453.6 \, gr}{1 lb} = ??$

Now we cancel the units: anything seen on top will cancel one set like it on the bottom: the lbs on top will be cancelled (erased) by the lbs on the bottom. Now we multiply 3 X 453.6 to get 1360.8 on the top; the bottom will be 1 X 1 to get 1. The only units left are grams and they are on top which is where they need to be if we want grams; thus, the answer is 1360.8 grams.

The main point is to identify what is known, arrange the conversion factor such that units can be cancelled and the desired units are on top. Often times, several conversion factors are needed. We can put any number in there, as long as they are valid factors whereby the top is equivalent to the bottom. See the example right whereby we want to know how many ounces of coffee are in a 5 gallon pot:

$$\frac{5 \, gal}{1} \times \frac{4 \, qt}{1 \, gal} \times \frac{2 \, pt}{1 \, qt} \times \frac{2 \, cup}{1 \, pt} \times \frac{8 \, oz}{1 \, cup} = ??$$

After we do the unit cancellation and the multiplication, we get 640 oz.

Conversion factors can have numbers other than 1 on the lower half of the fraction; How many lbs in 3000 grams? After unit cancellation and the multiplication and division, we get 6.614 lbs after round off. Note: Always keep track of the units and perform all the necessary division.

$$\frac{3000 \, grams}{1} \times \frac{1 lb}{453.6 \, grams} = ??$$

See sample calculations below to get density & thermal conductivity converted to proper units for use in a cooling time calculation (see also pages on cooling time calculation).

$$0.90 \, \frac{gr}{cm^3} \times \frac{16.38 \, cm^3}{1 \, in^3} \times \frac{1728 \, in^3}{1 \, ft^3} \times \frac{1 lb}{453.6 \, gr} = 56.16 \, \frac{lb}{ft^3}$$

$$0.1255 \, \frac{W}{(m \cdot K)} \times \frac{0.5778 \, \frac{Btu}{hr \, ft \, °F}}{\frac{W}{(m \cdot K)}} = 0.0725 \, \frac{Btu}{hr \, ft \, °F}$$

Temperature Convert °F to °C (001-200 °F)

°F	°C	°F	°C	°F	°C	°F	°C
1	-17	51	11	101	38	151	66
2	-17	52	11	102	39	152	67
3	-16	53	12	103	39	153	67
4	-16	54	12	104	40	154	68
5	-15	55	13	105	41	155	68
6	-14	56	13	106	41	156	69
7	-14	57	14	107	42	157	69
8	-13	58	14	108	42	158	70
9	-13	59	15	109	43	159	71
10	-12	60	16	110	43	160	71
11	-12	61	16	111	44	161	72
12	-11	62	17	112	44	162	72
13	-11	63	17	113	45	163	73
14	-10	64	18	114	46	164	73
15	-9	65	18	115	46	165	74
16	-9	66	19	116	47	166	74
17	-8	67	19	117	47	167	75
18	-8	68	20	118	48	168	76
19	-7	69	21	119	48	169	76
20	-7	70	21	120	49	170	77
21	-6	71	22	121	49	171	77
22	-6	72	22	122	50	172	78
23	-5	73	23	123	51	173	78
24	-4	74	23	124	51	174	79
25	-4	75	24	125	52	175	79
26	-3	76	24	126	52	176	80
27	-3	77	25	127	53	177	81
28	-2	78	26	128	53	178	81
29	-2	79	26	129	54	179	82
30	-1	80	27	130	54	180	82
31	-1	81	27	131	55	181	83
32	0	82	28	132	56	182	83
33	1	83	28	133	56	183	84
34	1	84	29	134	57	184	84
35	2	85	29	135	57	185	85
36	2	86	30	136	58	186	86
37	3	87	31	137	58	187	86
38	3	88	31	138	59	188	87
39	4	89	32	139	59	189	87
40	4	90	32	140	60	190	88
41	5	91	33	141	61	191	88
42	6	92	33	142	61	192	89
43	6	93	34	143	62	193	89
44	7	94	34	144	62	194	90
45	7	95	35	145	63	195	91
46	8	96	36	146	63	196	91
47	8	97	36	147	64	197	92
48	9	98	37	148	64	198	92
49	9	99	37	149	65	199	93
50	10	100	38	150	66	200	93

Temperature Convert °F to °C (201-400 °F)

°F	°C	°F	°C	°F	°C	°F	°C
201	94	251	122	301	149	351	177
202	94	252	122	302	150	352	178
203	95	253	123	303	151	353	178
204	96	254	123	304	151	354	179
205	96	255	124	305	152	355	179
206	97	256	124	306	152	356	180
207	97	257	125	307	153	357	181
208	98	258	126	308	153	358	181
209	98	259	126	309	154	359	182
210	99	260	127	310	154	360	182
211	99	261	127	311	155	361	183
212	100	262	128	312	156	362	183
213	101	263	128	313	156	363	184
214	101	264	129	314	157	364	184
215	102	265	129	315	157	365	185
216	102	266	130	316	158	366	186
217	103	267	131	317	158	367	186
218	103	268	131	318	159	368	187
219	104	269	132	319	159	369	187
220	104	270	132	320	160	370	188
221	105	271	133	321	161	371	188
222	106	272	133	322	161	372	189
223	106	273	134	323	162	373	189
224	107	274	134	324	162	374	190
225	107	275	135	325	163	375	191
226	108	276	136	326	163	376	191
227	108	277	136	327	164	377	192
228	109	278	137	328	164	378	192
229	109	279	137	329	165	379	193
230	110	280	138	330	166	380	193
231	111	281	138	331	166	381	194
232	111	282	139	332	167	382	194
233	112	283	139	333	167	383	195
234	112	284	140	334	168	384	196
235	113	285	141	335	168	385	196
236	113	286	141	336	169	386	197
237	114	287	142	337	169	387	197
238	114	288	142	338	170	388	198
239	115	289	143	339	171	389	198
240	116	290	143	340	171	390	199
241	116	291	144	341	172	391	199
242	117	292	144	342	172	392	200
243	117	293	145	343	173	393	201
244	118	294	146	344	173	394	201
245	118	295	146	345	174	395	202
246	119	296	147	346	174	396	202
247	119	297	147	347	175	397	203
248	120	298	148	348	176	398	203
249	121	299	148	349	176	399	204
250	121	300	149	350	177	400	204

Temperature Convert °F to °C (401-600 °F)

°F	°C	°F	°C	°F	°C	°F	°C
401	205	451	233	501	261	551	288
402	206	452	233	502	261	552	289
403	206	453	234	503	262	553	289
404	207	454	234	504	262	554	290
405	207	455	235	505	263	555	291
406	208	456	236	506	263	556	291
407	208	457	236	507	264	557	292
408	209	458	237	508	264	558	292
409	209	459	237	509	265	559	293
410	210	460	238	510	266	560	293
411	211	461	238	511	266	561	294
412	211	462	239	512	267	562	294
413	212	463	239	513	267	563	295
414	212	464	240	514	268	564	296
415	213	465	241	515	268	565	296
416	213	466	241	516	269	566	297
417	214	467	242	517	269	567	297
418	214	468	242	518	270	568	298
419	215	469	243	519	271	569	298
420	216	470	243	520	271	570	299
421	216	471	244	521	272	571	299
422	217	472	244	522	272	572	300
423	217	473	245	523	273	573	301
424	218	474	246	524	273	574	301
425	218	475	246	525	274	575	302
426	219	476	247	526	274	576	302
427	219	477	247	527	275	577	303
428	220	478	248	528	276	578	303
429	221	479	248	529	276	579	304
430	221	480	249	530	277	580	304
431	222	481	249	531	277	581	305
432	222	482	250	532	278	582	306
433	223	483	251	533	278	583	306
434	223	484	251	534	279	584	307
435	224	485	252	535	279	585	307
436	224	486	252	536	280	586	308
437	225	487	253	537	281	587	308
438	226	488	253	538	281	588	309
439	226	489	254	539	282	589	309
440	227	490	254	540	282	590	310
441	227	491	255	541	283	591	311
442	228	492	256	542	283	592	311
443	228	493	256	543	284	593	312
444	229	494	257	544	284	594	312
445	229	495	257	545	285	595	313
446	230	496	258	546	286	596	313
447	231	497	258	547	286	597	314
448	231	498	259	548	287	598	314
449	232	499	259	549	287	599	315
450	232	500	260	550	288	600	316

Temperature Convert °F to °C (601-800 °F)

°F	°C	°F	°C	°F	°C	°F	°C
601	316	651	344	701	372	751	399
602	317	652	344	702	372	752	400
603	317	653	345	703	373	753	401
604	318	654	346	704	373	754	401
605	318	655	346	705	374	755	402
606	319	656	347	706	374	756	402
607	319	657	347	707	375	757	403
608	320	658	348	708	376	758	403
609	321	659	348	709	376	759	404
610	321	660	349	710	377	760	404
611	322	661	349	711	377	761	405
612	322	662	350	712	378	762	406
613	323	663	351	713	378	763	406
614	323	664	351	714	379	764	407
615	324	665	352	715	379	765	407
616	324	666	352	716	380	766	408
617	325	667	353	717	381	767	408
618	326	668	353	718	381	768	409
619	326	669	354	719	382	769	409
620	327	670	354	720	382	770	410
621	327	671	355	721	383	771	411
622	328	672	356	722	383	772	411
623	328	673	356	723	384	773	412
624	329	674	357	724	384	774	412
625	329	675	357	725	385	775	413
626	330	676	358	726	386	776	413
627	331	677	358	727	386	777	414
628	331	678	359	728	387	778	414
629	332	679	359	729	387	779	415
630	332	680	360	730	388	780	416
631	333	681	361	731	388	781	416
632	333	682	361	732	389	782	417
633	334	683	362	733	389	783	417
634	334	684	362	734	390	784	418
635	335	685	363	735	391	785	418
636	336	686	363	736	391	786	419
637	336	687	364	737	392	787	419
638	337	688	364	738	392	788	420
639	337	689	365	739	393	789	421
640	338	690	366	740	393	790	421
641	338	691	366	741	394	791	422
642	339	692	367	742	394	792	422
643	339	693	367	743	395	793	423
644	340	694	368	744	396	794	423
645	341	695	368	745	396	795	424
646	341	696	369	746	397	796	424
647	342	697	369	747	397	797	425
648	342	698	370	748	398	798	426
649	343	699	371	749	398	799	426
650	343	700	371	750	399	800	427

BAR to PSI Conversions

psi	bar	psi	bar	psi	bar	psi	bar
15	1	740	51	1465	101	2190	151
29	2	754	52	1479	102	2204	152
44	3	769	53	1494	103	2219	153
58	4	783	54	1508	104	2233	154
73	5	798	55	1523	105	2248	155
87	6	812	56	1537	106	2262	156
102	7	827	57	1552	107	2277	157
116	8	841	58	1566	108	2291	158
131	9	856	59	1581	109	2306	159
145	10	870	60	1595	110	2320	160
160	11	885	61	1610	111	2335	161
174	12	899	62	1624	112	2349	162
189	13	914	63	1639	113	2364	163
203	14	928	64	1653	114	2378	164
218	15	943	65	1668	115	2393	165
232	16	957	66	1682	116	2407	166
247	17	972	67	1697	117	2422	167
261	18	986	68	1711	118	2436	168
276	19	1001	69	1726	119	2451	169
290	20	1015	70	1740	120	2465	170
305	21	1030	71	1755	121	2480	171
319	22	1044	72	1769	122	2494	172
334	23	1059	73	1784	123	2509	173
348	24	1073	74	1798	124	2523	174
363	25	1088	75	1813	125	2538	175
377	26	1102	76	1827	126	2552	176
392	27	1117	77	1842	127	2567	177
406	28	1131	78	1856	128	2581	178
421	29	1146	79	1871	129	2596	179
435	30	1160	80	1885	130	2610	180
450	31	1175	81	1900	131	2625	181
464	32	1189	82	1914	132	2639	182
479	33	1204	83	1929	133	2654	183
493	34	1218	84	1943	134	2668	184
508	35	1233	85	1958	135	2683	185
522	36	1247	86	1972	136	2697	186
537	37	1262	87	1987	137	2712	187
551	38	1276	88	2001	138	2726	188
566	39	1291	89	2016	139	2741	189
580	40	1305	90	2030	140	2755	190
595	41	1320	91	2045	141	2770	191
609	42	1334	92	2059	142	2784	192
624	43	1349	93	2074	143	2799	193
638	44	1363	94	2088	144	2813	194
653	45	1378	95	2103	145	2828	195
667	46	1392	96	2117	146	2842	196
682	47	1407	97	2132	147	2857	197
696	48	1421	98	2146	148	2871	198
711	49	1436	99	2161	149	2886	199
725	50	1450	100	2175	150	2900	200

NOTE: Numbers rounded to nearest whole number;
1 bar = 14.504 psi; 1 kg/cm^2 = 14.223 psi

Calculate Area - Common Shapes

Rectangle = A×B

Parallelogram = B×H

Triangle = ½ B×H

Right Triangle = ½ A×B

Trapezoid = ½ H×(B$_1$+B$_2$)

Hexagon = 0.866×D^2

Octagon = 0.828×D^2

Segment = area sector − area triangle

¼ Round = 0.7854×R^2

90° Fillet = 0.215×R^2

Ellipse = 0.7854×A×B

Parabolic Section = 2/3 XY

Circle = 0.7854×D^2 = (π×D^2)/4

Annulus = 0.7854×(D^2− d^2)

Sector = (R^2×A)/115
where A is in degrees

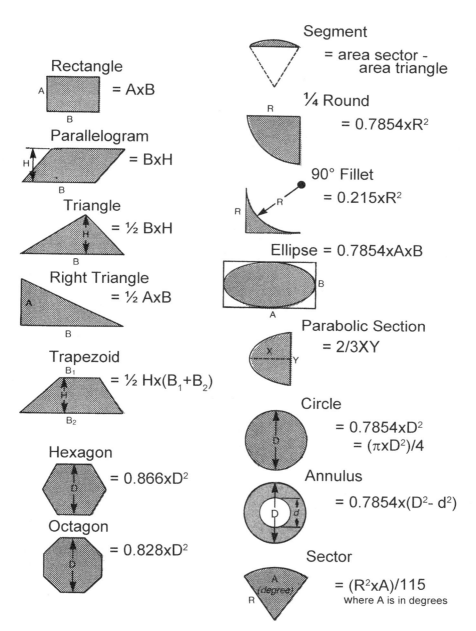

Millivolt Output For Type J T/C's (0 - 310 °F)

TYPE J THERMOCOUPLE OUTPUT IN MILLIVOLTS FOR DIFFERENT TEMPERATURES. NOTE: REFERENCE JUNCTION @ 32 DEGREES (ZERO mv); SUBTRACT VALUE FOR AMBIENT
If you start with Mv then ADD Mv for Ambient temp: Sum equals actual temp in Mv........If you start with Temp then convert to Mv then subtract Mv for ambient (should equal actual Mv measured)

°F	0	1	2	3	4	5	6	7	8	9	10
0	-0.885	-0.858	-0.831	-0.803	-0.776	-0.748	-0.721	-0.694	-0.666	-0.639	-0.611
10	-0.611	-0.583	-0.556	-0.528	-0.501	-0.473	-0.445	-0.418	-0.390	-0.362	-0.334
20	-0.334	-0.307	-0.279	-0.251	-0.223	-0.195	-0.168	-0.140	-0.112	-0.084	-0.056
30	-0.056	-0.028	0.000	0.028	0.056	0.084	0.112	0.140	0.168	0.196	0.224
40	0.224	0.253	0.281	0.309	0.337	0.365	0.394	0.422	0.450	0.478	0.507
50	0.507	0.535	0.563	0.592	0.620	0.648	0.677	0.705	0.734	0.762	0.791
60	0.791	0.819	0.848	0.876	0.905	0.933	0.962	0.990	1.019	1.048	1.076
70	1.076	1.105	1.134	1.162	1.191	1.220	1.248	1.277	1.306	1.335	1.363
80	1.363	1.392	1.421	1.450	1.479	1.507	1.536	1.565	1.594	1.623	1.652
90	1.652	1.681	1.710	1.739	1.768	1.797	1.826	1.855	1.884	1.913	1.942
100	1.942	1.971	2.000	2.029	2.058	2.088	2.117	2.146	2.175	2.204	2.233
110	2.233	2.263	2.292	2.210	2.350	2.380	2.409	2.438	2.467	2.497	2.526
120	2.526	2.555	2.585	2.614	2.644	2.673	2.702	2.732	2.761	2.791	2.820
130	2.820	2.849	2.879	2.908	2.938	2.967	2.997	3.026	3.056	3.085	3.115
140	3.115	3.145	3.174	3.204	3.233	3.263	3.293	3.322	3.352	3.381	3.411
150	3.411	3.441	3.470	3.500	3.530	3.560	3.589	3.619	3.649	3.678	3.708
160	3.708	3.738	3.768	3.798	3.827	3.857	3.887	3.917	3.947	3.976	4.006
170	4.006	4.036	4.066	4.096	4.126	4.156	4.186	4.216	4.245	4.275	4.305
180	4.035	4.335	4.365	4.395	4.425	4.455	4.485	4.515	4.545	4.575	4.605
190	4.605	4.635	4.665	4.695	4.725	4.755	4.786	4.816	4.846	4.876	4.906
200	4.906	4.936	4.966	4.996	5.026	5.057	5.087	5.117	5.147	5.177	5.207
210	5.207	5.238	5.268	5.298	5.328	5.358	5.389	5.419	5.449	5.479	5.509
220	5.509	5.540	5.570	5.600	5.630	5.661	5.691	5.721	5.752	5.782	5.812
230	5.812	5.843	5.873	5.903	5.934	5.964	5.994	6.025	6.055	6.085	6.116
240	6.116	6.146	6.176	6.207	6.237	6.268	6.298	6.328	6.359	6.389	6.420
250	6.420	6.450	6.481	6.511	6.541	6.572	6.602	6.633	6.663	6.694	6.724
260	6.724	6.755	6.785	6.816	6.846	6.877	6.907	6.938	6.968	6.999	7.029
270	7.029	7.060	7.090	7.121	7.151	7.182	7.212	7.243	7.274	7.304	7.335
280	7.335	7.365	7.396	7.426	7.457	7.488	7.518	7.549	7.579	7.610	7.641
290	7.641	7.671	7.702	7.732	7.763	7.794	7.824	7.855	7.885	7.916	7.947
300	7.947	7.977	8.008	8.039	8.069	8.100	8.131	8.161	8.192	8.223	8.253

Millivolt Output For Type J T/C's (310 - 610 °F)

TYPE J THERMOCOUPLE OUTPUT IN MILLIVOLTS FOR DIFFERENT TEMPERATURES. NOTE: REFERENCE JUNCTION @ 32 DEGREES (ZERO mv); SUBTRACT VALUE FOR AMBIENT
If you start with Mv then ADD Mv for Ambient temp; Sum equals actual temp in Mv........If you start with Temp then convert to Mv then subtract Mv for ambient (should equal actual Mv measured)

°F	0	1	2	3	4	5	6	7	8	9	10
310	8.253	8.284	8.315	8.345	8.376	8.407	8.437	8.468	8.499	8.530	8.560
320	8.560	8.591	8.622	8.652	8.683	8.714	8.745	8.775	8.806	8.837	8.867
330	8.867	8.898	8.929	8.960	8.990	9.021	9.052	9.083	9.113	9.144	9.175
340	9.175	9.206	9.236	9.267	9.298	9.329	9.359	9.390	9.421	9.452	9.483
350	9.483	9.513	9.544	9.575	9.606	9.636	9.667	9.698	9.729	9.760	9.790
360	9.790	9.821	9.852	9.883	9.914	9.944	9.975	10.006	10.037	10.068	10.098
370	10.098	10.129	10.160	10.191	10.222	10.252	10.283	10.314	10.345	10.376	10.407
380	10.407	10.437	10.468	10.499	10.530	10.561	10.592	10.622	10.653	10.684	10.715
390	10.715	10.746	10.777	10.807	10.838	10.869	10.900	10.931	10.962	10.992	11.023
400	11.023	11.054	11.085	11.116	11.147	11.177	11.208	11.239	11.270	11.301	11.332
410	11.332	11.363	11.393	11.424	11.455	11.486	11.517	11.548	11.578	11.609	11.640
420	11.640	11.671	11.702	11.733	11.764	11.794	11.825	11.856	11.887	11.918	11.949
430	11.949	11.980	12.010	12.041	12.072	12.103	12.134	12.165	12.196	12.226	12.257
440	12.257	12.288	12.319	12.350	12.381	12.411	12.442	12.473	12.504	12.535	12.566
450	12.566	12.597	12.627	12.658	12.689	12.720	12.751	12.782	12.813	12.843	12.874
460	12.874	12.905	12.936	12.967	12.998	13.029	13.059	13.090	13.121	13.152	13.183
470	13.183	13.214	13.244	13.275	13.306	13.337	13.368	13.399	13.430	13.460	13.491
480	13.491	13.522	13.553	13.584	13.615	13.645	13.676	13.707	13.738	13.769	13.800
490	13.800	13.830	13.861	13.892	13.923	13.954	13.985	14.015	14.046	14.077	14.108
500	14.108	14.139	14.170	14.200	14.231	14.262	14.293	14.324	14.355	14.385	14.416
510	14.416	14.447	14.478	14.509	14.539	14.570	14.601	14.632	14.663	14.694	14.724
520	14.724	14.755	14.786	14.817	14.848	14.878	14.909	14.940	14.971	15.002	15.032
530	15.032	15.063	15.094	15.125	15.156	15.186	15.217	15.248	15.279	15.310	15.340
540	15.340	15.371	15.402	15.433	15.464	15.494	15.525	15.556	15.587	15.617	15.648
550	15.648	15.679	15.710	15.741	15.771	15.802	15.833	15.864	15.894	15.925	15.956
560	15.956	15.987	16.018	16.048	16.079	16.110	16.141	16.171	16.202	16.233	16.264
570	16.264	16.294	16.325	16.356	16.387	16.417	16.448	16.479	16.510	16.540	16.571
580	16.571	16.602	16.633	16.663	16.694	16.725	16.756	16.786	16.817	16.848	16.879
590	16.879	16.909	16.940	16.971	17.001	17.032	17.063	17.094	17.124	17.155	17.186
600	17.186	17.217	17.247	17.278	17.309	17.339	17.370	17.401	17.432	17.462	17.493

Ohm's Law For D.C. Circuits

P = Power in watts
I = Current in Amperes
E = Voltage
R = Resistance in Ohms

$P = EI$ $\qquad$ $R = \dfrac{E}{I}$

$P = I^2 R$ $\qquad$ $R = \dfrac{E^2}{P}$

$P = \dfrac{E^2}{R}$ $\qquad$ $R = \dfrac{P}{I^2}$

$I = \dfrac{E}{R}$ $\qquad$ $E = \dfrac{P}{I}$

$I = \dfrac{P}{E}$ $\qquad$ $E = IR$

$I = \sqrt{\dfrac{P}{R}}$ $\qquad$ $E = \sqrt{PR}$

Misc. Electrical Formulas

$$\text{ohms} = \frac{\text{volts}}{\text{amps}} \qquad \text{amps} = \frac{\text{volts}}{\text{ohms}} \qquad \text{volts} = \text{amps} \times \text{ohms}$$

Power in AC Circuits:

$$\text{Efficiency} = \frac{746 \times \text{output horsepower}}{\text{input watts}}$$

$$3\Phi \text{ KW} = \frac{\text{volts} \times \text{amps} \times \text{pf} \times 1.732}{1000}$$

$$3\Phi \text{ volt amps} = \text{volts} \times \text{amps} \times 1.732$$

$$3\Phi \text{ efficiency} = \frac{746 \times \text{horsepower}}{\text{volts} \times \text{amps} \times \text{pf} \times 1.732}$$

$$3\Phi \text{ pf} = \frac{\text{input watts}}{\text{volts} \times \text{amps} \times 1.732}$$

$$1\Phi \text{ amps} = \frac{746 \times \text{horsepower}}{\text{volts} \times \text{efficiency} \times \text{pf}}$$

$$1\Phi \text{ efficiency} = \frac{746 \times \text{horsepower}}{\text{volts} \times \text{amps} \times \text{pf}}$$

$$1\Phi \text{ pf} = \frac{\text{input watts}}{\text{volts} \times \text{amps}}$$

$$3\Phi \text{ HP} = \frac{\text{volts} \times \text{amps} \times \text{pf} \times 1.732 \times \text{eff}}{746}$$

$$1\Phi \text{ HP} = \frac{\text{volts} \times \text{amps} \times \text{pf} \times \text{eff}}{746}$$

Full Load Transformer Current

3φ VOLTAGE LINE TO LINE

KVA RATING	208	230	460	600	4160
3	8.3	7.5	3.8	2.9	0.4
6	16.7	15.1	7.5	5.8	0.8
9	25.0	22.6	11.3	8.7	1.2
15	41.6	37.7	18.8	14.4	2.1
30	83	75	38	29	4
45	125	113	56	43	6
75	208	188	94	72	10
100	278	251	126	96	14
150	416	377	188	144	21
225	625	565	282	217	31
300	833	753	377	289	42
500	1388	1255	628	481	69
750	2082	1883	941	722	104
1000	2776	2510	1255	962	139
1500	4164	3765	1883	1443	208
2000	5552	5021	2510	1925	278
2500	6940	6276	3138	2406	347
5000	13879	12551	6276	4811	694
7500	20819	18827	9414	7217	1041
10000	27758	25103	12551	9623	1388

Formula used: AMPS = (KVA x 1000) / (VOLTS x 1.732)

1φ VOLTAGE

RATING	120	208	230	460	600
1	8.3	4.8	4.3	2.2	1.7
3	25.0	14.4	13.0	6.5	5.0
5	41.7	24.0	21.7	10.9	8.3
7.5	62.5	36.1	32.6	16.3	12.5
10	83.3	48.1	43.5	21.7	16.7
15	125.0	72.1	65.2	32.6	25.0
25	208	120	109	54	42
37.5	313	180	163	82	63
50	417	240	217	109	83
75	625	361	326	163	125
100	833	481	435	217	167
125	1042	601	543	272	208
167.5	1396	805	728	364	279
200	1667	962	870	435	333
250	2083	1202	1087	543	417
333	2775	1601	1448	724	555
500	4167	2404	2174	1087	833

Formula used: AMPS = (KVA x 1000) / VOLTS

Full Load Motor Current

3φ A.C. INDUCTION TYPE VOLTAGE			
HP	208	230	460
1/2	2.2	2.0	1.0
3/4	3.1	2.8	1.4
1	4.0	3.6	1.8
1.5	5.8	5.2	2.6
2	7.5	6.8	3.4
3	10.6	9.6	4.8
5	16.8	15.2	7.6
7.5	24	22	11
10	31	28	14
15	46	42	21
20	60	54	27
25	75	68	34
30	88	80	40
40	115	104	52
50	143	130	65
60	170	154	77
75	212	192	96
100	273	248	124
125	344	312	156
150	396	360	180
200	528	480	240

1φ VOLTAGE		
HP	115	230
1/6	4.4	2.2
1/4	5.8	2.9
1/3	7.2	3.6
1/2	9.8	4.9
3/4	13.8	6.9
1	16	8
1.5	20	10
2	24	12
3	34	17
5	56	28
7.5	80	40
10	100	50

Elements Of A Molding Cycle

A typical injection molding cycle consists of the following main components:

COOLING: Packing pressure
Hold pressure
<u>Cooling time and plasticizing time</u>

OPEN TIME: <u>Mold open</u>
Core pull and/or eject
Mold open dwell (stop time)
Core set if applicable
<u>Mold close</u> and clamp tonnage buildup

INJECTION: Core set if applicable sequence
Injection delay
Actual fill time

The above is listed in typical order of time required, whereby the cooling and/or plasticizing require the most time. Items underlined typically offer the most opportunity for improvement.

How to Reduce the Cycle

PRODUCT DESIGN CONSIDERATIONS:
Minimize wall thickness; Adequate sidewall draft (1/2 deg +/side); Increased radii in corners.

IMPROVED MOLD DESIGN:
Provide generous amount of cooling channels; Calculate coolant flow and Reynolds number; Select mold steel by reviewing and balancing thermal conductivity with tendency of steel to result in water line corrosion which causes heat transfer barriers and blockages. Stainless steels have a lower thermal conductivity, but generally are a good choice as water treatment programs rarely are adequate to prevent water line corrosion. Provide generous amount of ejection so parts can be ejected without excessive part distortion. Provide adequate venting so faster fill speeds can be used without burning. Do not oversize the gate; start small and increase if necessary.

MOLD SETUP IN PRESS:
Select free flowing water line sizes and fittings; quick disconnects should not be shut off type as they typically restrict flow. Bypasses or loops should be minimized. Do not put all the "INS" on one side and "OUTS" on opposite side; alternate directions to better reduce temperature differential. Use flow meter/indicators to verify presence of flow. If the temperature differential between supply and return is greater than 4 degrees (°F), the flow rate is probably poor and inadequate; good flows typically have as low as 2° F temperature rise. Optimize mold fast close, slowdown, mold open, and other control settings. Most mold setups can be improved by fine tuning switch positions. Avoid excessive mold open dwell time. Eject on the fly (during mold open) if your machine can do so. Set mold and melt temperatures as low as practical; faster injection speeds can still improve flow even at low end melt temperatures. Fine tuning to reduce cycle by 1/2 second is worth the effort.

NOTE:
Faster cycles will typically result in higher shrink rates; thus, cycle optimization needs to be identified during the mold prototyping or development stages.

Calculating Plastic Machine Load

$$\text{Load} = \frac{\text{Vol (annual)} \times \text{Cycle (sec)} \times (1 + \text{rej \%})}{(\text{Mach Util \%}) \times \text{\# cavs} \times 240 \times 24 \times 3600}$$

Comments & assumptions:
Machine utilization is often est @ 90% (0.90)
cavs is the mold cavitation
240 is typical number of working weekdays per year (usually 240-250 w/o holidays)
24 hrs for a working day (if 3 shift operation)
3600 seconds in an hour
Reject rate must be estimated (5% is often used: 0.05)

A load of 1.0 is a full load working 5 days per week.
A load of 1.2 indicates a full load for 6 days per week.
A load of 1.4 indicates 7 days per week is required.

Planning for a load of 1.0 - 1.2 is desirable to get good machine utilization, but yet allows time to make up for unforeseen problems or schedule increases.

In order to calculate hours required for a given short run:

$$\text{Hours} = \frac{\text{Volume} \times \text{Cycle (sec)} \times (1+\text{rej \%})}{3600 \times \text{\# cavs}}$$

The aforementioned formulas can easily be written into spreadsheet type software to keep track of the total load on a given molding press. This type of forward planning is useful to predict if and when overtime will be required or when machine purchases are needed. The load on molds should also be tracked, so time is available for preventive maintenance.

Hot Runner vs. Cold Runner

Advantages of Hot Runner Molds
- Less pressure drop thru runner
- Potential for reduced gate vestige without trimming
- Less or zero runner regrind
- Potential for faster cycle (less fill required & ease of ejection - less mold open time; cycle not limited by runner cooling)
- May be required for large cavitation (full hot or semi-hot)
- Full hot runner has no runner to separate - no separation losses
- Less projected area resulting in reduced clamp tonnage required
- Permits top center gating without use of three plate system (or other gate locations which may be difficult in cold runner)

Disadvantages of Hot Runner Molds
- Initial cost and complexity (mold and controller)
- Possible increased difficulty if performing frequent color changes
- Increased mold maintenance required
- Possible increased difficulty to start-up
- Increased residence time at melt temperature (possible thermal degradation)

Hot Runner Startup & Operation

Hot runner systems offer the following benefits to the molder: elimination or reduction of regrind; automation of part handling is easier; potential for faster cycles; part/runner separation is avoided; injection pressures may be lower. It is important to follow certain guidelines to get the most out of your hot runner system.

Some mold makers will list min and max temperature differentials for the manifold system versus the mold temperature; these should be followed as they allow for proper preload from thermal expansion: This is important to prevent leakage and/or excessive compression damage. A heat soak time of approximately thirty minutes should be allowed after set points are reached to permit proper thermal expansion (assumes resin w/ good thermal stability ... some will require less heat soak time). Front clamping plate cooling is necessary to prevent excessive heating of the platens. It is worthwhile to start-up the system with all zones off; then check each zone by itself to determine that the T/C response is for the respective zone heating. Always make a cavity sheet which indicates which zone number is what cavity and what section of mold relates to which manifold zone. Set the manifold temperature equal to the process melt temperature. Adjust drops as needed to achieve proper gate vestige. Chilled molds need to be started at room temperature. Do not connect gate coolant water in series. If your controller doesn't have soft start or ground fault capabilities; set the temps at only one hundred degrees for the first half hour so as to dry the absorbed moisture out of heater insulation. If a filter nozzle is used, great care is needed in filter nozzle selection to avoid excessive pressure drops - both on injection and ability to decompress manifold before mold opens. Always perform a balance of fill analysis before the mold leaves the mold maker; insist on a maximum imbalance of ten to twenty percent (depending on cavitation). It is best to never purge through the open hot runner mold due to potential for leakage and potential damage to cavities (i.e. some systems which have multiple cavities in a single cavity block could develop forces greater than retention screws retaining cavities or plate strength). It is necessary to sometimes purge through an open mold to purge degraded resin: try to purge with back pressure or minimal injection pressures; never use operating pressures to purge through the open mold. Never connect/disconnect plugs which are carrying power as it can damage most controllers besides being unsafe.

Injection Molding Press Selection

The following parameters should be reviewed when selecting or comparing presses:
1. Dry cycle time - indicator of clamp speed.
2. Injection rate capability - speed.
3. Screw recovery rate (calculate needed rate).
4. Clamp tonnage and type.
5. Tie bar diameter, tie bar clearances for mold.
6. Platen size, thickness, support to ways, etc.
7. Screw and barrel L/D, bi-metallic, tip type.
8. Oil capacity: multi-pumps or accumulators - can different machine functions occur at same time, such as clamp open and ejection or core pull.
9. Injection speed profile capabilities.
10. Method of transfer from fill to hold psi.
11. Min/max daylight (opening) of clamp.
12. Max resin pressure vs hyd pressure req'd.
13. SPI KO patterns present.
14. If toggle - location of force to platen.
15. Temperature control - # of zones, PID type?
16. Multi-ejection; dwell forward; intermediate stop/dwell capability: useful for part/ runner separation if mold is so designed.
17. Platen cooling - useful if warm/hot molds are run; platen thermal expansion results in tie bar movement which requires more mold closing and mold protection pressures (also wear issue).
18. Air eject circuits are useful inexpensive options - provide air assisted part blow off.
19. Cavity vent circuits can provide air blast just prior to and during mold open to bias part sticking to proper mold half - requires related mold design features.
20. Shut-off nozzles can be useful for many molds.
21. Ejection force or tonnage should be reviewed.
22. Injection carriage pull-in force can also be an issue if cylinder nozzles are used (control is useful).
23. Review inj carriage length ability to extend beyond locating ring for molds with deep recesses.

When selecting a press: size platens so two thirds of tie bar space (horiz & vert) is used by the mold. Select shot sizes such that at least forty percent of shot will be used each cycle; this reduces thermal degradation as well as providing more control of the injected shot - remember that machine controls are on the linear movement of the screw. Do not exceed eighty percent. Most machine deficiencies are poor plasticizing capability and/or lack of oil volume to perform multiple functions at same time and achieve fast clamp speeds or injection speeds. Other common deficiencies are long injection/plasticizing strokes to achieve a requested shot size - this may result in the screw controlling the cycle.

Checking Platens For Squareness

Parallelism of platens is important for several reasons:
A. Ease of platen movement with optimum alignment; reduced closing pressures required
B. Required for even loading of mold faces; misalignment would result in misaligned mold halves and flash.
C. Misalignment creates uneven loading of tie bars.

Toggle presses should have the parallelism checked at least twice per year; hydraulic presses at once a year. Re level press before checking parallelism. Alignment and parallelism is checked via the following:
1. If platen support ways exist, the lower tie bars should be parallel to the ways: check with a gaging parallel across tie bars; measure depth to the way from the parallel (parallel within 0.004 inches).
2. Fixed platen should be square to 2 lower tie bars (within 0.003 inches). Check via gaging V-block with perpendicular indicator stand and two dial indicators. Place V-block over tie bar and slide toward platen until indicators move; record differential of movement. Note: zero calibrate indicators using large tool makers square and surface plate. Check moving platen in same manner.
3. Platen to platen parallelism can now be checked at this point with an ID mic and extensions. Platens should be parallel to within 0.004 inches.

The nozzle centering should also be checked to the locating ring: check at four places; adjust to within 0.005 inches.

Tie bar stretch should also be checked to insure equal tie bar loading.

Injection Molding Press Preventive Maintenance

DAILY:
1. Clean periphery of the machine (floor & press).
2. Check hydraulic oil level for proper level & temp.
3. Check central lube system level; fill if needed.
4. Check lube drain catch basins; clean if needed.
5. Check nitrogen psi (if accumulator is present).
6. Check heat exchanger for adequate coolant flow.
7. Check mold clamps for tightness.
8. Check for excessive resin leakage at nozzle.
9. Check front gate safety - drop bar set correct.
10. Does front gate opening stop the press closing.
11. Check barrel zones for set point being achieved.
12. Check feed throat water for adequate cooling.
13. Listen for excessive pump noise.

WEEKLY:
14. Clean all Y-strainers supplying machine & equipment.
15. Check heater bands for proper operation.
16. Check all safety devices.

EACH 3 MONTHS:
17. Apply grease to grease nipples (list locations).
18. Clean oil suction filter.
19. Examine barrel heaters and terminals.
20. Examine barrel heater thermocouples[1].
21. Inject into blockage to test check valve[2].

ANNUAL:
22. Clean heat exchanger.
23. Check condition of hydraulic oil (visual & lab).
24. Clean or replace oil filtration element.
25. Re-level press.
26. Check platen squareness & parallelism.
27. Realign nozzle to locating ring (if needed).
28. Clean lube oil reservoir.
29. Remove screw; measure screw and barrel[3].
30. Clean feed throat cooling.
31. Examine linkage pins if toggle.
32. Check for equal tie bar strain.
33. Verify capability of specified pressures.
34. Check for loose elect. cards, terminals, etc.

---

[1] Lightly burnish T/C contact surfaces with abrasive paper.
[2] Set inj psi @ 40%; inject into previously filled mold for five sec; observe screw position.
[3] Measure and record dimensions at room temperature.

Injection Molding Safety

Safety is everyone's responsibility in the workplace. Safety is most often related to good maintenance and good housekeeping. Safety needs to be an attitude that is always present in your daily activities. Employees should not be hesitant to voice safety concerns in the workplace. Management is just as committed to safety as the operators on the floor; the primary difference is that the operators are usually the closest to unsafe conditions; keep management advised of unsafe conditions.

The following includes items to be maintained to assure a safe working environment:
1. Floor and machine should be kept free of oil.
2. Floor and machine should be kept free of pellets.
3. Never reach over or under machine guards.
4. Never climb between tie bars when pumps are running.
5. Retract inj unit before entering tie bar space.
6. The front gate should have an electrical, hydraulic and mechanical safety device preventing clamp from closing when the front gate is open.
7. The rear gate should have an electrical interlock preventing clamp from closing when rear gate is open (there is often a hydraulic interlock here also).
8. Re-adjust mechanical safety each time mold open daylight space is adjusted.
9. The purge shield should prevent injection forward if the purge shield limit switch is not made.
10. Catwalks or platforms with railing should be present if hoppers such as drying hoppers stand tall enough whereby access requires climbing onto machine.
11. Know location of portable fire extinguishers; there should be an extinguisher no farther than 75 feet.
12. All electrical outlets should be marked as to voltage.
13. Never reach into the throat of an operating granulator. Unplug granulators before working on.
14. Always wear suitable foot and eye protection; safety glasses should be worn and steel toed shoes are recommended; soft soled shoes should not be worn.
15. Do not operate any equipment unless suitable training has been supplied.
16. All employees should be advised of any chemicals in the facility which are considered hazardous; read "Right To Know" laws for each particular state.
17. First aid kits should be available.
18. Advise operators that injection molding resin pressure can reach 30,000 psi and that hydraulic line pressure can reach 2500 psi. Clamp tonnage developed equals 2000 lbs of force for each ton; operators be advised.
19. Be conscious of sharp square corners on ejector pins; many cuts result from protruding ejector pins.
20. Razor knives also require extreme caution as their use results in many cuts.
21. NEVER use steel tools on the mold cores, cavities or parting line ... use brass, copper or aluminum. Brass can scratch highly polished steel, so use caution.
22. Do not stick fingers or rods into the barrel/screw feed throat area.
23. Examine air hoses and electrical cords to verify condition is proper; do not use cords with damaged insulation. Be especially observant when working near nozzle heater bands as these wires are easily damaged.
24. Use only swivel type safety eyebolts; screw eyebolts far enough in such that thread engagement is 1.5 times the diameter.
25. Never stand directly below a mold suspended in air.
26. Avoid back injuries; lift properly with the back upright and straight; know your limitations and do not exceed; use proper tools and get help when needed.

Screw Recovery vs Temperature

Hotter rear zone temperatures may yield faster screw recovery with some resins!

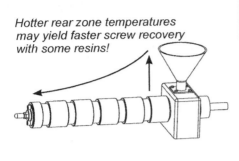

REAR ZONE TEMP (°F)	SCREW RECOVERY (sec)
430	7.95
440	7.49
450	7.07
460	6.66
470	6.41
480	6.32
490	6.28
500	6.31
510	6.40
520	6.51
530	6.67

NOTES:
1. Front zone (and hot runner manifold) should be set same as actual melt temperature.
2. Descending profiles may be beneficial to achieve faster screw recovery.

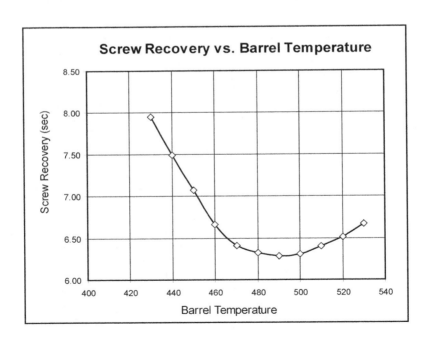

Injection Screw Terminology

<u>L/D</u> - Flighted screw length divided by screw diameter
<u>Compression</u> - Feed flight depth divided by metering depth; assuming constant pitch. Suggested compression ratios (low = 1.4-2.4, med = 2.4-3.0, high = 3.0-3.5)

LOW ... ABS (2.0); ASA (2.0); CA, CAB & CAP (2.2); PBT (1.8-2.0); PS & HIPS (2.0); PVC - flexible (1.4-2.0 w/ long compression zone @ 40% of length); PVC - rigid (1.6-2.2 w/ long comp. zone @ 40% of length); SAN (2.0); LDPE & HDPE (2.0-2.4); PP (2.0-2.4); PPS (2.3); PSU (2.1)

MED ... PC (2.6); PMMA (2.8); POM (2.5); PEI (2.5); PEEK (2.6); PET (2.4); PPO-Noryl (2.5);

HIGH ... EVA (3.0); PA (3.2)

<u>Feed Section</u> - Hopper end of screw where resin enters; this section has the deepest flights; the feed section is typically half the screw length.
<u>Transition Section</u> - Transition from feed to metering; in this section the screw flight depth is changing which accomplishes the compression.
<u>Metering Section</u> - The exit end of screw; usually 1/4 of the total screw length; the flight depth here is the smallest of the screw
<u>Bi-Metallic</u> - A second harder/more durable surface which is welded onto screw flights.
<u>Nitrided Screw</u> - Surface hardness through heat treatment in a high nitrogen atmosphere; hardness depth is approximately 0.020 - 0.024 inches.
<u>Barrier Screw</u> - Refers to the presence of barrier flights usually starting where the transition section begins. A barrier flight is a second flight located between normal flights. Barrier flights do not stand as tall as normal flights; this allows melt to travel over tops of flights and keep unmelted solids back until melting is accomplished.
<u>Check Valves</u> - Usually sliding ring type or ball check type; each acts as a one way valve allowing melt to convey forward during plasticizing but preventing back flow during injection. Ball checks usually accomplish better shut off, but ring types are more streamlined with less ΔP and less degradation to heat sensitive resins.
<u>General Purpose Screw</u> - A screw which usually is 2.5 or 2.6:1 compression and L/D of 20:1.
<u>Clearance</u> - 0.001 TO 0.0015 inches per side per inch of barrel diameter (e.g. A 2" barrel ID would have a total diametral clearance of 0.004 to 0.006 inches).

Simplified formula for estimating screw output:

$$\text{Output (lbs/hr)} = 2.3 \times D^2 \times h \times g \times N$$

D is screw diameter (inches),
h is metering depth (inches),
g is resin specific gravity at melt temp,
N is screw RPM

Calculating Color Blend Ratios

Color concentrates are developed by suppliers to a specific color as requested by the molder. The blend ratio is determined by the concentrate supplier. The following formulas can be used to calculate concentrate required:

Examples below use a 25:1 (Resin -"R" : color - "C") ratio: (e.g. .....for example)

A: Starting with given amount of color concentrate:
Use resin equal to "R" times the weight of conc: e.g. 40 lbs of conc is available; weigh 25 X 40 or 1000 lbs of resin; combine with the 40 lbs conc and blend.

B: Starting with a given amount of resin:
Add color concentrate equal to resin wt divided by "R": e.g. 300 lbs of resin is available; divide 300 by 25 to get 12; add 12 lbs of color to the 300 lbs resin and blend.

C: Blending to achieve a specified total blend weight:
Divide end total weight by ("R"+"C"); this will be the concentrate amount; add resin equal to "R" times the concentrate: e.g. A container holding 300 lbs is to be filled to capacity; 300 divided by 26 determines the amount of color at 11.54 lbs; combine with 11.54 X 25 or 288.5 lbs of resin (The exact total is slightly off 300 due to round off).

D.   Blending at the press with screw auger feeders:
Weigh total shot weight and divide by 26; each molding shot or cycle should allow this amount of color to fall. It usually is necessary to shut off the resin hopper to isolate the color from resin to gather a sample of color. Ideally the color is metered out simultaneous to the screw rotation.

NOTES REGARDING PERCENTAGES:

1. When blending to a total weight as in example "C" above, 25:1 is not 96% resin and 4% color, but rather 25/26 or 96.15% resin and 1/26 or 3.85% color.
2. In example "B" above where we start with only the resin or the 300 lbs, 25:1 resin/color could be re-stated as 1:25 color/resin which equals 1/25 or 0.04 or 4% times the resin as the color added to the resin (e.g. 300 X 0.04 = 12 lbs color added, but the 12 lbs color is 3.85% of the total of 312 lbs, 12/312 = 0.0385 or 3.85% of the total blend).
3. If the color is specified as 4% instead of a ratio, <u>we must assume we are blending to the total blend weight</u> whereby we would multiply the total times 0.04, (e.g. 300 lbs total X 0.04 = 12 lbs which leaves 288 lbs of resin; 12+288=300). The ratio here is 96/100:4/100 which reduces to 24:1 (not 25:1 as in earlier examples).

Color concentrate is usually considerably higher in cost than the resin; thus, use only the necessary amount. NOTE: Don't worry about differences above between 25:1 vs 24:1 or 3.85% vs 4.00 % above ... listed to best explain most accurate math ... start with what supplier provides as a blend ratio, then use best fit example above.

Resin Drying Temperatures

MATERIAL	TEMP (°F)	TIME (hrs)	BULK DENS. (lb/ft³)
ABS	190°	3-4	42
ACETAL (HOMOPOLYMER)	200°	1-2	40
ACETAL (COPOLYMER)	210°	1-2	40
ACRYLIC	180°	2-3	42
CELLULOSE BUTYRATE	170°	2-3	39
CELLULOSE ACETATE	170°	2-3	38
CELLULOSE PROPIONATE	170°	2-3	40
IONOMER (SURLYN)	160°	7-8	44
LCP (XYDAR)	300°	3-4	50
NYLON	180°	4-5	41
POLYCARBONATE	250°	3-4	40
P'CARB/PBT/ELAST (XENOY)	260°	3-4	42
PEEK	310°	3-4	52
PET-BOTTLE (EASTAPAK 9921)*	340°	5-6	52
POLYESTER (RYNITE)	275°	3-4	54
PET (THERMX EG001)	265°	4	42
P'ESTER (PBT/PET VALOX 815)	360°	4-5	48
POLYESTER (PBT-VALOX 420)	260°	2-3	48
POLYESTER (PET-VALOX 700)	320°	4-5	48
PETG (EASTAR GNXXX)	155°	6	46
PCTG (EASTAR DNXXX)	165°	6	45
PCTG/P'CARB (EASTALLOY DAXXX)	200°	4-6	43
PCT (THERMX CG907)	160°	4-6	41
PCT (THERMX CG921)	160°	6	50
PCTA (THERMX AG230)	330°	4	44
POLYARYLATE	250°	5-6	50
POLYETHERIMIDE	310°	4-5	52
POLYETHYLENE (BLACK)	160°	3-4	34
PPO/STYRENE (NORYL)	210°	2-3	49
POLYPHENYLENE SULFIDE	280°	2-3	50
POLYSULFONE	275°	3-4	50
POLYURETHANE	180°	2-3	48
SAN	180°	3-4	40
SMA (DYLARK)	200°	2-3	38
TPE (HYTREL)	210°	2-3	48
TPR (SANTOPRENE)	160°	2-3	46

* Can dry crystallized PET bottle resin at lower temps for longer time (i.e. 8 hrs @ 295° F).

NOTES:
1. Process return air should be cooled to 150° F or below. The desiccant will be more efficient with increased affinity for moisture at temps below 150°.
2. Fines have a higher ratio of surface area to mass; thus, will absorb moisture faster and more readily than pellets.
3. Some dryers require as much as 4 hours to regenerate; thus, to be considered when dryer is first turned on. Cool down of the bed regeneration should be with previously dried process air so as not to load the desiccant with moisture.
4. Calculate and allow for proper residence time.
5. Clean filters on a regular basis.
6. Excessive drying may result in color shift (some resins, especially clear).

Calculating Clamp Tonnage Required

The injection unit on a machine performs a pressure amplification: 2,000 psi hydraulic oil pressure may create 20,000 - 50,000 psi pressure on the plastic leaving the nozzle. This is done by intensification or mechanical advantage:

$$P1 \times A1 = P2 \times A2$$

P1 = Plastic Pressure exiting the nozzle .... A1 = Area of the Screw Diameter
P2 = Hydraulic Oil Injection Pressure .... A2 = Total Area of Inject Cylinder

In the past, most mechanical advantages were 10:1, but this is not true today. It is common to find values ranging from 10:1 to 16:1, but values up to 25:1 do exist. The variation is because some manufacturers may vary screw/barrel sizes to offer different shot sizes. Molders should know what each machine's mechanical advantage (intensification ratio) is for each press, especially if molds are run in different presses each time.

Once this amplified pressure is injected into the mold, it causes a second amplification to occur whereby it's pressure, which is PSI acting in all direction, is applied to the projected area of the mold cavities. This simple math formula below indicates what force is trying to open the mold halves:

$$\frac{pounds}{in^2} \times in^2 \text{ projected area} = \text{pounds of force}$$

The clamp tonnage must be large enough to oppose this force trying to blow the mold open. The tonnage is just that ... 2000 lbs of force for each ton (assuming U.S. tons; metric tons are 10 % greater - 2200 lbs of force for each metric ton).

EXAMPLE

Molded part = 2" X 3"
There are 4 cavs
Total projected area = 24 in^2

Screw diam = 50 mm or use 1.9685"
Inj ram diam = 7.5" (press spec as built)
Pack psi = 1000 (process setpoint)
Plastic psi = 14,516 (calculated)
Max hyd psi = 2000 (press spec as built)
Max inj psi = 29,032 (2000x44.178/3.0434)
Mech Adv = 14.516:1 (A2/A1 or 44.178/3.0434)
Clamp tons = 200 (press spec as built)

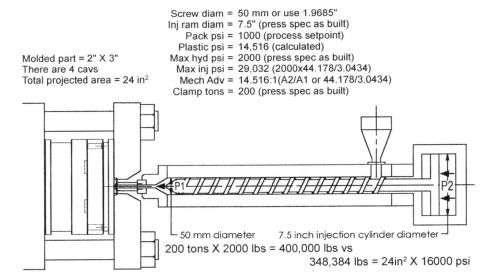

50 mm diameter   7.5 inch injection cylinder diameter
200 tons X 2000 lbs = 400,000 lbs vs
348,384 lbs = 24in^2 X 16000 psi

NOTE: The 348,384 is without pressure drop, which could be as high as 50% (minimal for nylon); thus, the formula as it appears, is conservative ... but in this case there is ample clamp tonnage to keep mold closed during fill/pack!

Setting and Starting The Mold

MOLD SETUP:
1. Determine resin requirements & availability; clean hopper, magnet, loader; load resin; start dryer if required.
2. Locate proper KO bars - all having equal length.
3. Prepare platens & mold using stone and mineral spirits.
4. Select eyebolt hole which yields a level hang/lift.
5. Do NOT stand below hanging mold; avoid hitting tie bars when lowering mold into place.
6. Line up locating rings; slowly close mold.
7. Level mold if not already and clamp to fixed platen.
8. Open moving platen (w/ hoist still attached/supporting mold); install KO bars. If KOs are acting as pullbacks: tighten bars making certain they bottom out against the ejector plate in mold.
9. Close platen; clamp mold to moving platen; remove safety straps; unhook hoist.
10. Open mold to desired daylight; set slowdown switches with certainty that banging the mold will not occur; fine tune the final switch positions by repetitive small adjustments, observations and readjustment.
11. Secure KOs to ejector plate of press; set stroke.
12. Connect all required power - hyd, electric, pneumatic.
13. Make sure powered functions are functional; run electrical heaters just long enough to prove functionality avoiding excessive heat buildup before water is connected.
14. Connect water lines using an acceptable number of loops/jumpers; locate lines clear of any interference. Avoid having all "INS" on the same side.
15. Recheck fittings for proper connection; turn water on; (heaters should be off); look for leaks.

PROCESS SETUP (IF UNKNOWN):
16. Set barrel profile per resin supplier's recommended mid-range (same logic for mold temp).
17. Estimate the shot size and set machine for approximately 2/3 of the mold's full shot. Set decompression stroke. Set a position transfer point (if machine is so equipped) approximately one inch from bottom.
18. Estimate & set second stage time; set second stage pressure at zero.
19. Set 1st stage pressure at 50% for starters (this may ultimately be set at 100% - assuming Decoupled Molding^{SM}).
20. Set velocity to maximum.
21. Estimate and set cooling time.
22. Set back pressure at 50 psi.
23. Refer also to Hot Runner Start-up procedure if applicable.

MOLD START-UP:
24. Purge barrel free of degraded resin.
25. Set machine for semi-auto; start cycle; observe screw.
26. Adjust velocity and/or pressure as needed; if the fill was fast and short as estimated, the pressure can be increased. The fill pressure should be set high enough so the fill speed is not pressure limited, but controlled by velocity setpoints. If flash or dieseling occur, slow the velocity.
27. After observing each cycle, the shot size and transfer point will be adjusted frequently to set the process so that the first stage accomplishes 95 - 98 % of the fill as measured by shot weight.
28. Once the first stage shot size, transfer, velocity and pressure are set, we can set 2nd stage packing pressure.
29. Adjust pack pressure as needed, but do not overpack.
30. Recheck cushion; some cushion should be maintained.
31. Set screw rpm so recovery is completed just prior to next cycle, but not limiting cycle time.

PROCESS DOCUMENTATION:
32. Record all basic machines setpoints on the setup sheet.
33. Note the transfer time (fill time) and weight.
34. Note the overall cycle time.
35. Note the ejection: multiple, push only, push/pull, etc.
36. Total shot weight, part weight, % runner, etc.

IMPORTANT NOTE: When using Decoupled Molding^{SM} (see also pages discussing this technique): Considerable skill and specific mold and machine knowledge is required when setting pressures near maximum. Set pressures in accordance with consideration for mold damage in the event some parts do not shoot due to gate blockage and remaining cavities actually see the elevated 1st stage pressure.

Determining Gate Seal Time

Knowledge of the gate seal time, permits the molder to use the maximum (if needed) injection forward time to accomplish effective packing. Injection forward times less than the gate seal time often result in sinks, voids and increased shrinkage resulting in poor dimensional compliance. The total injection forward time should be set a fraction longer than the gate seal time if cycle time permits. Hot tip gates may not result in a seal due to localized packing around the hot gate. This packing is of little or no benefit; thus, do not attempt a full gate seal on some hot gates parts. Note also: valve gates are sealed mechanically... this test applies differently to valve gated molds; whereas, it can still indicate optimum injection time, but valve must be closed in time to achieve good gate vestige.

Hold Time (sec)	Shot Weight (parts only) (grams)
0.60	8.716
0.70	8.770
0.80	8.810
0.90	8.830
1.00	8.850
1.10	8.863
1.20	8.878
1.30	8.885
1.50	8.902
1.70	8.913
1.80	8.915
2.00	8.916
2.20	8.916
2.40	8.916

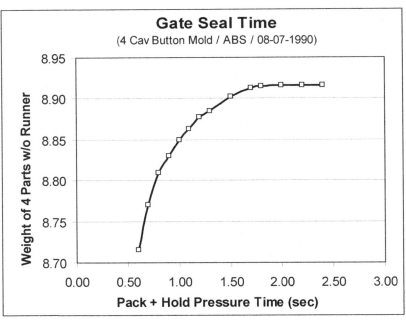

Gate Seal Time
(4 Cav Button Mold / ABS / 08-07-1990)

Balance of Fill Analysis

Cavity	Avg Part Wt (grams)	Fill Seq	Avg % Unbal	% Full	Part wts (grams)		
					shot 1	shot 2	shot 3
4	1.40	1	0.00%	100.00%	1.42	1.39	1.38
5	1.39	2	0.72%	99.28%	1.39	1.39	1.38
12	1.38	3	1.19%	98.81%	1.39	1.37	1.38
13	1.38	4	1.43%	98.57%	1.37	1.39	1.37
6	1.37	5	1.91%	98.09%	1.38	1.37	1.36
14	1.34	6	3.82%	96.18%	1.34	1.35	1.34
11	1.33	7	4.77%	95.23%	1.36	1.32	1.31
3	1.33	8	5.01%	94.99%	1.35	1.32	1.31
15	1.19	9	14.56%	85.44%	1.20	1.21	1.17
10	1.15	10	17.66%	82.34%	1.16	1.15	1.14
9	1.14	11	18.38%	81.62%	1.15	1.15	1.12
7	1.12	12	19.57%	80.43%	1.12	1.14	1.11
1	1.11	13	20.29%	79.71%	1.13	1.11	1.10
8	1.06	14	23.87%	76.13%	1.10	1.05	1.04
16	1.06	15	23.87%	76.13%	1.07	1.06	1.06
2	1.03	16	26.01%	73.99%	1.04	1.03	1.03

PROCEDURE:
1. Process should be running at equilibrium with regards to mold & melt temperature, pressures normal & stable.
2. Consider mold's ability to run a short shot.
3. Set feed or transfer to run short shots whereby one part is full (or nearly so); begin shot collection.
4. Collect three shots separated by cavity number, weigh individual parts and record, calculate average.
5. Arrange data in table, sorted in descending order (from high to low) of average part weight.
6. Compute % unbalanced (this mold has a 26% imbalance - see table above).
7. Restore process to normal.
8. Graph serves no purpose other than display of data.

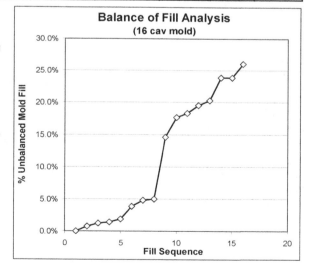

Balance of Fill Analysis (16 cav mold)

Mechanics of Polymer Flow

The term "non-newtonian" is the term used to describe polymer flow. Newtonian flow behavior means that the viscosity does not change with a change in shear rate (flow speed). Polymer flow, which is non-newtonian, does result in viscosity reduction as shear rate increases.

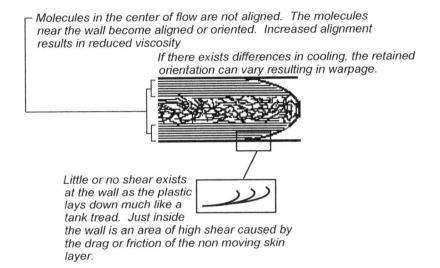

Molecules in the center of flow are not aligned. The molecules near the wall become aligned or oriented. Increased alignment results in reduced viscosity

If there exists differences in cooling, the retained orientation can vary resulting in warpage.

Little or no shear exists at the wall as the plastic lays down much like a tank tread. Just inside the wall is an area of high shear caused by the drag or friction of the non moving skin layer.

It is also important to understand what effect average molecular weight and weight distribution have on the polymer flow. A higher molecular weight indicates longer molecules which will result in stiffer flow (and enhanced physical properties). The weight is described as an average. The weight distribution describes the amount of long and short molecules. A broad distribution means abundance of a wide variety of lengths. Narrow versus wide distribution can be compared to a pitcher of ice water where the water represents short molecules and the ice represents long molecules. A pitcher of slush also has ice and water, but the proportions of each are different and the ice is much smaller; however, the average size could be the same. The slush could be thought of as a narrow distribution. The slush (or narrow distributions) would flow stiffer or more viscous than would the water with ice cubes. Some variation in molecular length is both inherent to resin production and is needed to some extent for injection molding. The short length molecules act as a carrier to aid in flow and resulting processability.

It is a challenge and often a compromise to determine exactly what average weight and distribution yields the optimum combination of physical properties and processability. The weight distribution is not easily controlled by the resin manufacturer; customers may receive variation from one lot to the next. It is often worthwhile to identify your resin's average weight and distribution when it is running good for future reference. Various labs can perform this testing.

Relative Viscosity Testing

Relative viscosity testing is done to determine how the resin's viscosity changes as injection speed increases. The theory behind using high speed injection rates is to take advantage of the decreased viscosity which results when plastics are subjected to high shear rates. The viscosity drops because the long chain molecules align themselves which reduces the viscosity.

As can be seen on the graph (couple pages forward), there is a region on the left side of graph where small changes in fill time/rate result in large changes in viscosity. These large changes in viscosity result in molding variation with regards to actual cavity pressure seen by the molded part. It can also be seen on the graph, that the right side - very fast fill rates - levels out nearly horizontal indicating that the viscosity is more stable. It can be concluded that the fast fill rates should be more forgiving with regards to inherent molding variation. Fill rates should not; however, be pushed to the point of melt fracture; whereby, molecular weight is reduced.

In order to take advantage of the machine's full capability however, the molder must set the molding process in what is called Decoupled Molding^{SM} (Service Mark of RJG Associates, Inc ... used with permission from Rodney Groleau, RJG Associates, Inc. Traverse City, MI). Decoupled Molding^{SM} separates the pressure used for filling from the pressure used for packing by transferring from fill to pack prior to complete mold fill. This transfer is usually done at approximately the 94-96% filled by weight point of filling. This is done because flash would likely result if the higher fill pressures (used for faster fill) were also used to pack the mold. Separating fill from pack yields improved control of the fill and allows more flexibility toward optimizing both the fill and pack of the mold.

When performing this test, adhere to the following:

1. The screw should not bottom out; if the screw bottoms out, a hydraulic fluid "hammering" effect will yield erroneous readings in the hydraulic pressure gauge.
2. Mold should not completely fill (near full on fastest fill).
3. It should be mentioned that the pressure cannot be captured fast enough on the mechanical-analog (dial) gauge as it does not respond fast enough to capture the peak (and is usually too far from the injection cylinder). A peak holding gauge from a hydraulic pressure transducer is required to capture short term peaks.
4. Record fill time from start of injection until the transfer from 1st stage to 2nd stage. The second stage pressure should be zeroed out as we are only concerned with the fill.
5. It is also necessary to know the machine's mechanical advantage (ratio of max plastic psi to max hydraulic pressure-avail from supplier specs). See sample calculations on the supplied table.
6. Record the actual melt temperature from which the test is run.
7. Run test with hydraulic pressure at maximum machine setting.
8. Start test at maximum fill rate, drop down in steps; perform approximately 10 different fill rates from fast to slow.

Sample of Relative Viscosity Data

SAMPLE DATA FORMAT AND CALCULATIONS FOR CALCULATING RELATIVE VISCOSITY

A VELOCITY SETPOINT (mm/sec)	B ACTUAL FILL TIME (sec)	C PEAK HYD PRESSURE (kg/cm^2)	D PEAK HYD PRESSURE (psi)	E MACHINE'S MECHANICAL ADVANTAGE	RELATIVE VISCOSITY (B*D*E)	1/FILL TIME
300	0.31	100.0	1422	15.43	6802	3.226
250	0.35	97.1	1381	15.43	7457	2.857
200	0.41	92.2	1311	15.43	8294	2.439
150	0.53	86.2	1226	15.43	10024	1.887
100	0.77	73.2	1041	15.43	12367	1.299
75	1.02	67.4	958	15.43	15084	0.980
50	1.51	57.3	815	15.43	18984	0.662
35	2.14	51.4	731	15.43	24135	0.467
20	3.71	42.6	606	15.43	34678	0.270

See graph on next page

Viscosity Formula

The following formulas are used to calculate viscosity. Relative viscosity calculations do not include flow path area; thus, it is relative to the mold and machine system including it's various orifices and wall thickness.

$$\text{viscosity} = \frac{\text{shear stress}}{\text{shear rate}} = \frac{\frac{\text{Force}}{\text{area}}}{\frac{\text{volumetric flow rate}}{\text{volume}}}$$

$$\text{shear stress} = \frac{\text{Force}}{\text{area}} = \frac{\text{pounds}}{\text{in}^2}$$

$$\text{shear rate} = \frac{\text{volumetric flow rate}}{\text{volume}} =$$

$$= \frac{\frac{\text{in}^3}{\text{sec}}}{\text{in}^3} = \frac{\text{in}^3}{\text{sec}} \times \frac{1}{\text{in}^3} = \frac{1}{\text{sec}} = \text{sec}^{-1}$$

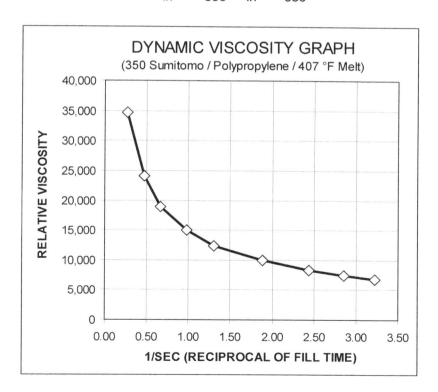

DYNAMIC VISCOSITY GRAPH
(350 Sumitomo / Polypropylene / 407 °F Melt)

Machine Load Sensitivity

A machine's load sensitivity is an indicator of it's repeatability under varying loads. Varying loads result as the resin viscosity varies due to varying melt temps and resin variation; mold temperature can also influence the force required to fill.

If a machine is operating to perfection there is no difference in the result of the load sensitivity calculation. A test of load sensitivity is done by injecting into the mold and comparing that data with data from air shot testing. A peak reading hydraulic gauge and a fill time clock are required to perform this test.

$$\text{Load sensitivity} = \frac{\frac{\text{time 1} - \text{time 2}}{\text{time 1}}}{\text{psi 1} - \text{psi 2}} \times 100{,}000$$

The answer reads % per 1000 psi
time 1 = fill time into the mold
time 2 = fill time of the air shot
psi 1 = the peak pressure into the mold
psi 2 = the peak pressure of the air shot

NOTE:
1. Do not let screw bottom out; we are only interested in the fill.
2. Late model machines may reduce the injection speed during air shots in a manual mode for safety reasons; if this occurs, the data will be incorrect. Such errors are readily seen on recorded strip chart plots of the fill.
3. See also rules for relative viscosity testing.

The average load sensitivity should be: less than 10 percent; 4% is good. See spreadsheet with real data on next page.

Machine Load Sensitivity Data Example

	VELOCITY SETPOINT (MM/SEC)	ACTUAL FILL TIME (SEC)	PEAK HYD PRESSURE (PSI)
INTO MOLD:	300	0.32	866
	250	0.38	807
	200	0.47	729
	150	0.63	650
	100	0.93	555
	75	1.24	494
	50	1.85	430
	35	2.64	382
	25	3.69	348
	15	6.12	308
	10	9.13	289
AIR SHOT:	300	0.32	601
	250	0.38	539
	200	0.47	518
	150	0.62	414
	100	0.93	338
	75	1.23	308
	50	1.84	261
	35	2.59	242
	25	3.65	176
	15	6.08	144
	10	9.06	123

VELOCITY SETPOINT	LOAD SENSITIVITY
300	0
250	0
200	0
150	6.71
100	0
75	4.33
50	3.19
35	13.49
25	6.32
15	3.99
10	4.63

13.49 IS MAX; 3.88 IS AVG

used with permission from Rodney Groleau, RJG Associates, Inc. Traverse City, MI

Sources of Variation

An injection molding operation has many sources of variation:
- Measurement error (operator or equipment)
- Core/cavity sizing differences (variability also exists in the mold making process) ... included in Machine (meaning Equipment) in diagram below
- Process variation inherent to the mold design (cooling, mold filling, packing, etc.)
- Process variation caused by molding press (screw & tip, hydraulics, controls, etc.)
- Process variation caused by aux. equipment (dryers, thermolators, blenders, etc.)
- Resin variation (molecular wt, weight distribution, dispersion of colorants or additives, etc.)
- Operators changing setup
- Post mold storage conditions (temperature, humidity, etc.)

Cause and Effect Diagrams ... aka ISHIKAWA or FISHBONE diagrams

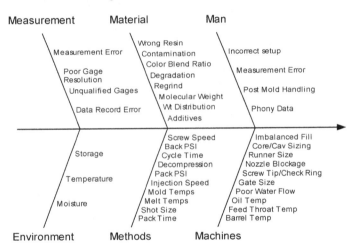

Sample of Histogram Usage

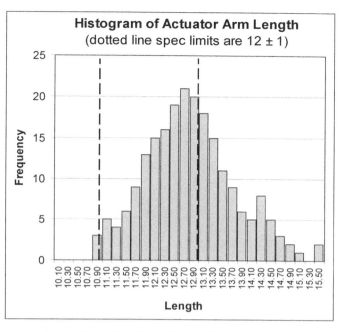

Avg Length	Min	Max	Freq
10.1	10.0	10.2	0
10.3	10.2	10.4	0
10.5	10.4	10.6	0
10.7	10.6	10.8	0
10.9	10.8	11.0	3
11.1	11.0	11.2	5
11.3	11.2	11.4	4
11.5	11.4	11.6	6
11.7	11.6	11.8	9
11.9	11.8	12.0	13
12.1	12.0	12.2	15
12.3	12.2	12.4	16
12.5	12.4	12.6	19
12.7	12.6	12.8	21
12.9	12.8	13.0	20
13.1	13.0	13.2	18
13.3	13.2	13.4	15
13.5	13.4	13.6	11
13.7	13.6	13.8	9
13.9	13.8	14.0	6
14.1	14.0	14.2	5
14.3	14.2	14.4	8
14.5	14.4	14.6	5
14.7	14.6	14.8	3
14.9	14.8	15.0	2
15.1	15.0	15.2	1
15.3	15.2	15.4	0
15.5	15.4	15.6	2

Sample Pareto Chart Data

Category	Total	Cat %	Cum %	CUM Defects
Gate Defect	288	41.1%	41.1%	288
"X" DIM	192	27.4%	68.5%	480
FM	87	12.4%	80.9%	567
Other	43	6.1%	87.0%	610
Broken	36	5.1%	92.2%	646
Color	23	3.3%	95.4%	669
Flash	21	3.0%	98.4%	690
Short Shot	7	1.0%	99.4%	697
"Z" DIM	2	0.3%	99.7%	699
"Y" DIM	2	0.3%	100.0%	701
	701	100%		

The pareto principle states that typically, 80% of the defects come from 20% of the causes. The following graph is a good way to graphically display where the problems are. Any problem solving venture needs to have the problem clearly identified.

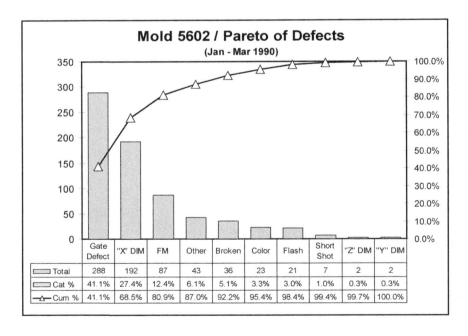

Examples of Process Capability

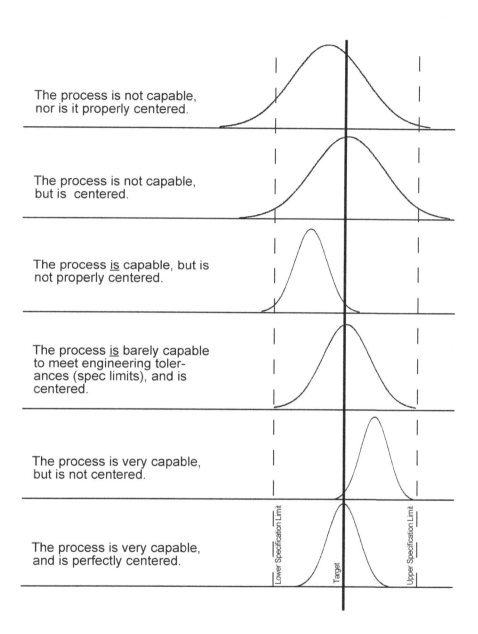

X Bar and R Control Charts

When the measured response is a quantitative variable (i.e. length, weight, etc) instead of an attribute (i.e. go, no-go), a X-bar & R chart is often used. Most charts used today can be purchased in a preformatted layout containing area for graphics, data logging, table of constants, orderly layout of data for sequential math operations, instructions, etc.

The mean or X-bar is an average of the samples; there are often 5 pieces in the sample (n=5). The grand mean or X-double bar is the average of the averages which is also the average of all measurements. The range or R is the spread within a sample of n pieces (high minus low). R-bar is the average of all the ranges. Sample sizes should be either 4 or 5. Wait until 20 samples of size n (n = 4 or 5) have been taken before calculating limits. Data must be time ordered on the chart – that is in chronological order of occurrence. Control limit formulas appear below the table on the next page. After control limits are calculated, the data points are either within the control limits (in control) or outside the control limits (out of control). "In control" data is considered random variation; out of control data is assignable to some specific special cause. Random data should only fall out outside the control limits three times in a thousand.

The main reason for keeping control charts are:
- Observe and record the patterns of process variation.
- Make adjustments to process before defects are produced.
- Use statistics to determine acceptable limits of variability.

Variation is inherent to manufacturing. Statistics can indicate when variation is within normal expected limits versus special causes which can be identified, corrected and possibly prevented from reoccurring by appropriate measures.

Variation comes from many sources besides your manufacturing process. Variation can occur in the raw material which resulted from the suppliers process. Variation occurs in the process of part measurement. Through careful engineering, studies can be made to quantify where the variation is and how much. Only by identification of the amount and source of variation can work be planned to reduce it. Do not underestimate the possible sources of variation. Fishbone diagrams are useful in identifying causes of variation.

Managers often call for SPC (Statistical Process Control) to be used in injection molding, but it is difficult to implement due to the time lag between: molding the parts; when the final shrinkage takes place; and the time required to measure and conclude where the process is with regards to dimensional compliance. One attempt to implement real time SPC is to perform a DOE (Design of Experiments) whereby planned process variation is introduced to the process. In a DOE with 4 factors (i.e. packing pressure, fill time, mold temperature and melt temperature), each at 2 levels, we would have an experiment with 16 different combinations of processes (assuming full replication). Such an experiment can be run in a shift; then critical dimensions are measured after a 48 hour shrinkage stabilization. If mold pressure transducers are used to record plastic pressure in the mold as one of the response variables, then correlation studies can be done on the other response variables which are less timely to capture during the actual molding process during production. During production the cavity pressure is tracked for SPC purposes. Some experimentation is required to determine optimum transducer location, but one successful choice has been to monitor (for SPC) near the last point of fill in the cavity and to control the process via transducers near the gate.

During new product development it can be useful to use statistical tolerancing so as to identify a design which has the potential for zero defects. Designed experiments are also highly recommended to determine the products sensitivity to process change. Statistical tolerancing and DOEs are beyond the scope of this pocket sized reference guide.

Example: X Bar and R Control Charts

HOUR	1	2	3	4	5	6	7	8	9	10	11	12	13	14	15	16	17	18	19	20	21	22	23	24
	76	74	77	79	75	74	72	77	75	76	74	73	77	72	74	76	76	73	71	78	79	74	73	76
	74	72	73	73	74	74	73	77	76	74	72	74	75	76	71	71	72	77	74	75	77	73	71	69
	69	71	78	74	74	73	77	71	74	79	72	73	76	71	77	76	78	72	74	74	72	77	65	76
	72	69	71	72	74	74	75	74	76	77	73	73	71	73	72	73	74	75	75	76	77	73	76	74
	74	74	75	76	76	74	73	73	71	71	77	71	75	77	74	75	73	77	77	75	72	71	71	71
Xbar =	73	72	74.8	74.8	74.8	74.6	73.8	74	74.4	75.4	73.6	72.8	74.8	73.8	73.6	74.2	74.6	74.8	74.2	75.6	75.4	73.6	71.2	73.2
Range =	7	5	7	7	2	1	5	6	5	8	5	3	6	6	6	5	6	5	6	4	7	6	11	7

$\overline{\overline{X}}$ = avg of $\overline{X}$s = Grand Avg or Grand mean = 74.025 = avg of all 120 "X" values

$\overline{R}$ = avg of Rs = 5.666

look up values for A_2, D_4 & D_3 below in appropriate table of constants

X bar chart; upper control limit (UCL) = $\overline{\overline{X}} + A_2\overline{R}$ = 77.29 (A_2 = 0.577 for n=5)

X bar chart; lower control limit (LCL) = $\overline{\overline{X}} - A_2\overline{R}$ = 70.76 (A_2 = 0.577 for n=5)

Range chart ... upper control limit (UCL) = $D_4\overline{R}$ = 11.98 (D_4 = 2.114 for n=5)

Range chart ... lower control limit (LCL) = $D_3\overline{R}$ = 0 (D_3 = 0 for n=5)

- D_3, D_4 & A_2 are found from a table of constants for variables control charts using n = 5 (5 ea hr).
- After comparing the actual values to the control limits, it can be seen the process is in control.
- It is recommended that the above data be plotted graphically to better display the data trends, etc.

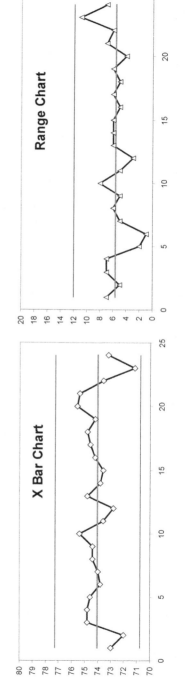

Tables of Constants for Control Charts

The following tables are for: Variables Data

SAMPLE SIZE n	Xbar-R CONTROL CHARTS			
	A_2	D_3	D_4	d_2
2	1.880	—	3.267	1.128
3	1.023	—	2.574	1.693
4	0.729	—	2.282	2.059
5	0.577	—	2.114	2.326
6	0.483	—	2.004	2.534
7	0.419	0.076	1.924	2.704
8	0.373	0.136	1.864	2.847
9	0.337	0.184	1.816	2.970
10	0.308	0.223	1.777	3.078

SAMPLE SIZE n	Xbar-s CONTROL CHARTS			
	A_3	B_3	B_4	c_4
2	2.659	—	3.267	0.7979
3	1.954	—	2.568	0.8862
4	1.628	—	2.266	0.9213
5	1.427	—	2.089	0.9400
6	1.287	0.030	1.970	0.9515
7	1.182	0.118	1.882	0.9594
8	1.099	0.185	1.815	0.9650
9	1.032	0.239	1.761	0.9693
10	0.975	0.284	1.716	0.9727

Process Capability

The following formulas can be used to compare your actual process performance to the specification limits. These formulas assume the process is in control and data is normally distributed about a central mean or average.

$$Cp = \frac{USL - LSL}{6\hat{\sigma}}$$

$\hat{\sigma}$ = estimate of process standard deviation = $\frac{\bar{R}}{d_2}$

When the Cp < 1 the process variation exceeds spec limits. When the Cp > 1 the process variation is less than spec. limits, but in either case the Cp only looks at process potential, meaning the actual mean must be centered within the spec limits to make full use of the limits with the process variation occurring.

It is stated above that Cp is process potential. Cpk is an indicator of process capability in that it looks at the variation relative to where the mean of the process is located.

$$Cpk\ (lower) = \frac{\bar{\bar{X}} - LSL}{3\hat{\sigma}} \qquad Cpk\ (upper) = \frac{USL - \bar{\bar{X}}}{3\hat{\sigma}}$$

Cpk = minimum of either {Cpk(lower), Cpk(upper)}.
Cpk values greater than 1 are favorable.
Ppk = minimum of either {Ppk(lower), Ppk(upper)} ... same as Cpk above except divisor is the calculated sample std dev (s) instead of estimated sigma as shown above.

Example:
Specification limit is 2.375 ± 0.010
The process grand average is 2.371
The r bar is 0.003; n = 4; thus, d2 is 2.09

$$\hat{\sigma} = \frac{\bar{R}}{d_2} = \frac{0.003}{2.059} = 0.001457$$

$$Cpk = \frac{\bar{\bar{X}} - LSL}{3\hat{\sigma}} = \frac{2.371 - 2.365}{3 \times 0.001457} = 1.37$$

The z distance can then be used to determine area Pz on the Z table to determine percentage of defects; it is calculated as follows. Note: The process must be in statistical control and the histogram should indicate a normal distribution

$$Z(lo) = \frac{\bar{\bar{X}} - LSL}{1\hat{\sigma}} \quad \& \quad Z(hi) = \frac{USL - \bar{\bar{X}}}{1\hat{\sigma}}$$

$$Z(lo) = \frac{2.371 - 2.365}{0.001457} = 4.11$$

4.11 yields zero defects (per table on next page, but 6 sigma tables would indicate a Z of 4.11 to be 21 ppm defects!), but if r bar had been 0.005 then the std dev would have been 0.002428 and Z would be 2.47 which would give an area Pz of 0.68% defects (0.0068). We should calculate Z(hi) and Ẑ(lo) in cases of large variation (range).

Z Table for Process Capability Calculations

PZ = THE PROPORTION OF PROCESS OUTPUT BEYOND A SINGLE SPECIFICATION LIMIT THAT IS Z STANDARD DEVIATIONS UNITS AWAY FROM THE PROCESS AVERAGE (FOR A PROCESS IN STATISTICAL CONTROL AND NORMALLY DISTRIBUTED).

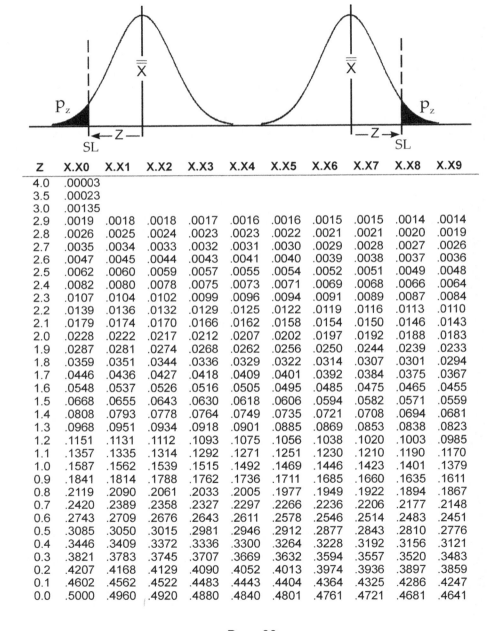

Z	X.X0	X.X1	X.X2	X.X3	X.X4	X.X5	X.X6	X.X7	X.X8	X.X9
4.0	.00003									
3.5	.00023									
3.0	.00135									
2.9	.0019	.0018	.0018	.0017	.0016	.0016	.0015	.0015	.0014	.0014
2.8	.0026	.0025	.0024	.0023	.0023	.0022	.0021	.0021	.0020	.0019
2.7	.0035	.0034	.0033	.0032	.0031	.0030	.0029	.0028	.0027	.0026
2.6	.0047	.0045	.0044	.0043	.0041	.0040	.0039	.0038	.0037	.0036
2.5	.0062	.0060	.0059	.0057	.0055	.0054	.0052	.0051	.0049	.0048
2.4	.0082	.0080	.0078	.0075	.0073	.0071	.0069	.0068	.0066	.0064
2.3	.0107	.0104	.0102	.0099	.0096	.0094	.0091	.0089	.0087	.0084
2.2	.0139	.0136	.0132	.0129	.0125	.0122	.0119	.0116	.0113	.0110
2.1	.0179	.0174	.0170	.0166	.0162	.0158	.0154	.0150	.0146	.0143
2.0	.0228	.0222	.0217	.0212	.0207	.0202	.0197	.0192	.0188	.0183
1.9	.0287	.0281	.0274	.0268	.0262	.0256	.0250	.0244	.0239	.0233
1.8	.0359	.0351	.0344	.0336	.0329	.0322	.0314	.0307	.0301	.0294
1.7	.0446	.0436	.0427	.0418	.0409	.0401	.0392	.0384	.0375	.0367
1.6	.0548	.0537	.0526	.0516	.0505	.0495	.0485	.0475	.0465	.0455
1.5	.0668	.0655	.0643	.0630	.0618	.0606	.0594	.0582	.0571	.0559
1.4	.0808	.0793	.0778	.0764	.0749	.0735	.0721	.0708	.0694	.0681
1.3	.0968	.0951	.0934	.0918	.0901	.0885	.0869	.0853	.0838	.0823
1.2	.1151	.1131	.1112	.1093	.1075	.1056	.1038	.1020	.1003	.0985
1.1	.1357	.1335	.1314	.1292	.1271	.1251	.1230	.1210	.1190	.1170
1.0	.1587	.1562	.1539	.1515	.1492	.1469	.1446	.1423	.1401	.1379
0.9	.1841	.1814	.1788	.1762	.1736	.1711	.1685	.1660	.1635	.1611
0.8	.2119	.2090	.2061	.2033	.2005	.1977	.1949	.1922	.1894	.1867
0.7	.2420	.2389	.2358	.2327	.2297	.2266	.2236	.2206	.2177	.2148
0.6	.2743	.2709	.2676	.2643	.2611	.2578	.2546	.2514	.2483	.2451
0.5	.3085	.3050	.3015	.2981	.2946	.2912	.2877	.2843	.2810	.2776
0.4	.3446	.3409	.3372	.3336	.3300	.3264	.3228	.3192	.3156	.3121
0.3	.3821	.3783	.3745	.3707	.3669	.3632	.3594	.3557	.3520	.3483
0.2	.4207	.4168	.4129	.4090	.4052	.4013	.3974	.3936	.3897	.3859
0.1	.4602	.4562	.4522	.4483	.4443	.4404	.4364	.4325	.4286	.4247
0.0	.5000	.4960	.4920	.4880	.4840	.4801	.4761	.4721	.4681	.4641

Standard Deviation, Mean & Normal Distribution

Mean or average = $\bar{X} = \dfrac{\Sigma X}{n}$

Sample standard deviation = $\sigma_{n-1} = \sqrt{\dfrac{\Sigma (X-\bar{X})^2}{n-1}}$

NOTE: The aforementioned standard deviation is for a <u>sample</u> from the population. The standard deviation σ for the population would use only n in the equation above (instead of n -1). Take care when using spreadsheet software for calculating standard deviations, so that the proper function is selected and used (std dev for sample or population). Note also that R bar ÷ d2 is an estimate of standard deviation (aka sigma hat). The standard deviation squared is known as the variance. Only the variance is additive when performing statistical tolerancing.

For normally distributed data, a bell shaped curve has data distributed as follows. A grand average (X double bar) is the central point of the curve:

$$\bar{\bar{X}} \pm 1\sigma = 68.26\%$$
$$\bar{\bar{X}} \pm 2\sigma = 95.44\%$$
$$\bar{\bar{X}} \pm 3\sigma = 99.73\%$$
$$\bar{\bar{X}} \pm 4\sigma = 99.994\%$$
$$\bar{\bar{X}} \pm 5\sigma = 99.99995\%$$
$$\bar{\bar{X}} \pm 6\sigma = 99.9999998\%$$

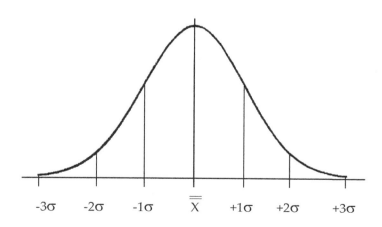

Tips, Tricks & Shortcuts

1. Check nozzle contact with sprue bushing using 0.002" brass shim stock; leaves good impression of contact area.
2. Check wall thickness in parts with pointed micrometers; saw parts so that the position of pieces does not become lost; mold component misalignment will result in thin walls (radiused corners, tapered walls).
3. Profiling the injection speed can be used to alter the fill pattern on imbalanced molds.
4. Never change gate size to balance fill.
5. Mold clamps should be long enough so that bolts are no more than halfway from mold to fulcrum point of clamp.
6. Leave screw forward on shutdown; otherwise start-up requires barrel to heat slug of plastic in front of screw, this incomplete melting is what often damages the check ring on subsequent start-up.
7. Watch screw during injection: observe speed, transfer, speed after transfer, ram bounce after transfer, lack of cushion, check amount of decompression, observe screw recovery; if possible look for screw rotation during injection.
8. Stop mold during opening before any ejection and look for signs of corner lifting, etc. which might indicate part trying to stick in stationary mold half.
9. Never assume water flow is present; verify via flow indicators. Even flow into a bucket can be misleading if a low ΔP exists between supply and return lines (checking water flow into bucket will not be pushing into normal back pressure on return line circuit). A large ΔT is usually a sure sign of low flow rates.
10. Check barrel heater band functionality often; new machines can be purchased with burnout alarms.
11. Always purge heat sensitive resins after idle time; especially on molds with tendency to stick.
12. Use as large of nozzle orifice as possible, but slightly smaller than sprue orifice.
13. Prior to shut down of a leaking hot runner system, change colors of resin to indicate location of leak (can feed several handfuls new color through magnet or feed throat); inject a few shots into mold after color change through nozzle is verified by air shots.
14. Excessive clamp tonnage on undersized molds can result in platen warp/wrap causing mold to flash.
15. Perform balance of fill analysis.
16. Determine gate seal time as a guide to set injection forward time.
17. Always establish and record the transfer weight and time if using decoupled molding.
18. When purging hygroscopic resins, observe air shot purgings for foaming; this indicates presence of moisture (assuming reasonable melt temperature).

Tips, Tricks & Shortcuts (ASCII & Symbol Codes)

19. Always cool clamping plates to prevent heating of platens; heated platens result in thermal expansion which results in tie bar movement effecting the clamp close and mold protection pressures required.
20. Never mix acetal and PVC ... HCl from over heated PVC + formaldehyde may result in an explosion.
21. Methyl gas formed by degraded polypropylene may attack and erode copper alloys.
22. Locate mold half alignment blocks on horizontal and vertical center lines; use parallel locks when vertical shutoffs exist (or near vertical). Tapered interlocks will not perform full alignment until mold is closed.
23. Color code water lines (i.e. blue "ins" and red "outs").
24. Remember the tangent value for an angle of 1 degree which is 0.01745. Can then multiply distance in inches x angle x 0.01745 = rise.

Useful ASCII code symbols (press ALT and at same time type numbers from numerical keypad on right side of keyboard) and use Symbol fonts.

Arial font (and others)		Symbol fonts	
ALT+	Character	typed	displayed
155	¢	s	σ
171	½	S	Σ
172	¼	p	π
174	«	D	Δ
175	»	r	ρ
230	µ	b	β
241	±	m	μ
246	÷	a	α
248	°	d	δ
249	•	f	ϕ
250	·	h	η
253	²	\	$\therefore$
		q	θ
		F	Φ
		W	Ω
		@	$\cong$
		^	$\perp$

Equipment Selection (Molding Press)

In order to prepare for any given molding project, there will be equipment required. The molding press is at the top of the list. Your mold will require certain accommodating features in order to even make the process doable. Many features are basic requirements and others can enhance processability if selection results in an optimum match.

Molding Press

1. SHOT SIZE – The advertised shot size of any molding press is with G.P. Polystyrene. To calculate the maximum shot capacity with your resin, perform the following calculation:

$$\frac{\text{sp grav your resin (gr/cc)}}{\text{sp grav P.S. (1.06 gr/cc)}} \times \text{Advertised shot size} = \text{Actual shot size}$$

Technically, the melt densities should be used in the above calculation, but room temperature density can be used in absence of melt density data. It is best not to exceed 80% of the maximum shot capacity when possible and also not to fall below 20% of capacity if possible. When using heat sensitive resins, it is best to calculate actual residence time at process temperatures (discussion in Troubleshooting Guide).

$$\frac{1.4 \times \text{bar cap (oz.)}}{\text{sp. grav PS (1.06)}} \times \frac{\text{sp grav PC (1.2)}}{\text{shot wt (oz.)}} \times \frac{\text{shot (oz.)}}{} \times \frac{1\text{min}}{60\text{ sec}} = \text{Time (min)}$$

Plan residence time to be between 1.5 and 4 minutes for optimum results (unless otherwise specified by resin supplier). Heat stable resins such as polyethylene and polypropylene can withstand higher residence times.

Equipment Selection (Mold Size vs Tie Bar Space)

2. MOLD SIZE VS. TIE BAR SPACE – Will the mold fit the press tie bar spacing. Most molds are installed thru the top (easier but not required); thus, the horizontal tie bar spacing must be checked to see if mold will fit. If not, check vertical tie bar clearance. This press has a 27.95" horizontal and vertical tie bar spacing – also known as a square pattern as opposed to a wide platen design whereby the horizontal clearance is wider than the vertical clearance. The K.O. patterns available are the 4" X 16" pattern (vertical and horizontal) and the 7" up/down and right/left patterns (and a center K.O. location). The next bigger S.P.I. standard K.O. pattern is 6" X 28" – not available on this press (then 10" X 40"). If only a center K.O. is used, the ejector plate must be thicker to endure usage without warpage and guided with pins/bushings to prevent binding. The widest possible pattern is normally preferred, in example above this is 4" X 16" pattern.

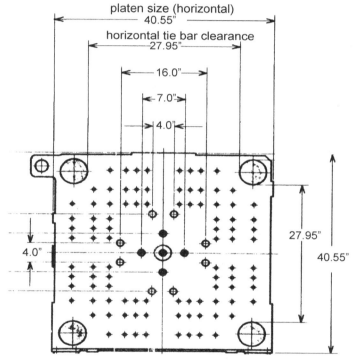

Equipment Selection (Tonnage & Injection Rate)

3. CLAMP TONNAGE – Compare projected area to clamp tonnage available. Often times the tons/in^2 are calculated and compared to "rules of thumb". These rules are very general guidelines and do not always work since resins come in a variety of melt index and molecular weight and weight distributions. This affects flow length and resulting injection pressure required. The part size and wall thickness also affect pressure required. As more hydraulic pressure is used, it is multiplied by the mechanical advantage and projected area of cavities (plus runner) to become the force trying to blow mold open resulting in flash. The tonnage is multiplied times 2000 pounds to get force holding mold closed. A simple math calculation of closing forces vs. opening forces is not accurate unless the pressure losses are known. It is sometimes suggested to use 40 - 60% as a multiplier to account for pressure drop. Tons/in^2 rules vary between 1 and 5 tons/in^2 depending on resin. Molds which have multiple parting lines such as stack molds and 3 plate molds typically use the opening with most projected area to calculate opening forces (regardless of calculation method). Resins which flash easy require more tons/in^2. Polystyrene often only requires 1 - 1½ tons/in^2, PP & PE may require 2 - 3 tons/in^2, polycarbonate may require 4 tons/in^2 and Nylon may require as much as 4 - 5. Accurate "rules of thumb" require experience with a particular resin and type of part design.

4. INJECTION RATE CAPABILITY VS. PLANNED FILL TIME – The machine will have a rated maximum injection rate in cc/sec (or some other units). The planned shot can be converted to volume in cc (cubic centimeters). If a shot is 250 grams of polycarbonate and the density is 1.20 gr/cc (1.02 gr/cc melt density) and we want a 0.7 second fill time; then the required fill rate performance for the machine must be at least 350 cc/sec. This is calculated as follows:

$$\frac{\text{injected shot weight (250 grams)}}{\text{sp grav @ melt (1.02 gr/cc)}} \Big/ \text{planned fill time (0.70 sec)} = 350 \text{ cc/sec}$$

The marketing literature supplied by the injection press manufacturer is often the best source to identify the maximum injection rate capability for the press.

5. MAXIMUM INJECTION PRESSURE – Injection pressure is a function of max hydraulic pressure and mechanical advantage. The press specification will normally list the maximum plastic pressure and max hydraulic pressure. The mechanical advantage is simply the plastic pressure divided by the hydraulic pressure. A given press size may have different mechanical advantages depending on screw size and injection unit ordered. With a given injection unit, as the screw diameter selected increases: so does shot size, injection rate and plasticizing rate capability, but maximum injection pressure will decrease. If maximum injection pressure is needed, then other features must be reduced based on the aforementioned relationship.

Equipment Selection (Plasticizing Rate)

6. PLASTICIZING RATE CAPABILITY VS. PLANNED COOLING TIME OR PLASTICIZING TIME – Many presses can only plasticize during cooling time because of hydraulic limitations. Presses equipped with suitable hydraulics and a nozzle valve may permit plasticizing simultaneous to mold open, eject and mold close operations. The calculation for plasticizing rate requirements by the molding press are similar to the injection rate calculation previously (in example calculation below the shot size will again be 250 grams of polycarbonate with a time available of 6 seconds; the press is rated at 244.8 Kg/hr which is converted to 68 grams/sec):

$$\frac{\text{shot size (250 grams)}}{\text{rated plasticizing rate (68 gr/sec) x 0.70}} = 5.25 \text{ seconds}$$

The rated plasticizing rate is multiplied times 0.70 to de-rate the unit (70% of rated or de-rated by 30%). The rated capacity is for polystyrene and we want to be sure the press has some margin of assurance that plasticizing can be accomplished; polycarbonate will likely not plasticize as easy as polystyrene. In this case, we calculate 5.25 seconds versus 6.0 seconds available; thus, no problem.

7. EJECTION K.O. PATTERN REQUIRED – Compare needed pattern to platen drawings; see also platen drawing on preceding pages.

Equipment Selection (Stroke & Daylight)

8. STROKE/DAYLIGHT REQUIREMENTS – On most machines we are concerned with the maximum daylight (maximum distance between stationary platen and moving platen) and the minimum shut height (minimum possible distance between two platens; thus, indicates the shortest mold which can be run in a press). These aforementioned terms however are not always used and clearly indicated by the press supplier. Toggle presses have an additional moving platen that the traditional moving platen strokes to and from. See the drawing below which lists the min and max stroke plus min and max mold size for this toggle press. Hydraulic presses do not have a minimum stroke requirement as do some toggle presses – this min stroke requirement is important to review for molds which have chains or stripper bolts which forever connect the two mold haves together during operation.

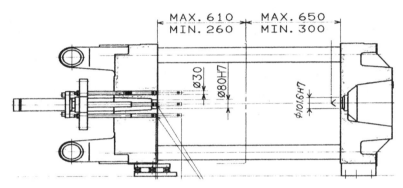

The maximum daylight on this press is the sum of max stroke and max mold height – 1260 mm. The minimum shut height is as indicated – 300 mm; maximum mold height is 650 mm (with hydraulic presses their is no max mold height specification; limitation is max daylight limits on ability to open mold and have room to eject parts with proper free fall). Not visible in drawing above is a maximum ejector stroke of 175 mm (6.88 inches). The locating ring is 101.6 mm or 4 inches.

Your mold requirements should be compared to this type scrutiny to verify mold can run as needed in the chosen press.

9. Other checks include availability of core pull circuits if needed (hydraulic and electrical); Multi-ejection stroke requirements; air blow off circuits if needed; hoist capacity on large molds as well as overhead clearance for setting the mold.

Auxiliary Molding Equipment - Dryer

1. DRYERS - The figure below is a Conair multi desiccant bed dehumidifying dryer. Important design and/or installation features for selection include:
   a. Multiple desiccant beds - 2 or more (dryer in figure has 4 beds).
   b. The regenerated bed is cooled down with previously dried air - not moist ambient air.
   c. The T/C controlling process air temperature circulated thru the hopper is located at the hopper air inlet - critical for optimum drying effectiveness.
   d. Aftercooler is needed if process temperature exceeds approximately 225° F (return air temperature entering current bed in process should be down to $\cong 150°$ F.)
   e. Locate hopper on molding press just above feed throat and magnet if possible. Size hopper to accomplish needed drying residence time.

Conair Dehumidifying Dryer
(graphic courtesy of Conair Franklin)

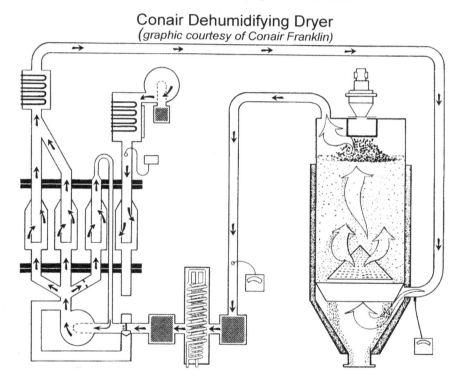

Conair Franklin, Route 8 North, Franklin, PA 16323 814-437-6861

Auxiliary Molding Equipment - MTCU

2. MTCU – MOLD TEMPERATURE CONTROL UNITS – Often abbreviated as MTCU. These are often called "thermolators", but Thermolator® is a registered trademark name for a MTCU marketed by Conair. The Conair Thermolator® is also a very good piece of equipment for circulating water thru a mold at a closely controlled temperature. MTCUs were first used primarily to heat up the water for molds running resins such as polycarbonate which needed warm or hot molds for proper processing. Now many molders use MTCUs for molds running at cold OR hot temperatures. The use of a MTCU often provides closer temperature control than achieved by a central or portable chilled water system; thus, reduced variability for the molded product.
In addition to the close temperature control, the coolant flow rate is boosted by the MTCU's circulating pump. It should be noted that MTCUs only heat the water; thus, the controlled temperature will only be at or above the water supplied to the MTCU. This water can be from a cooling tower or a chiller.

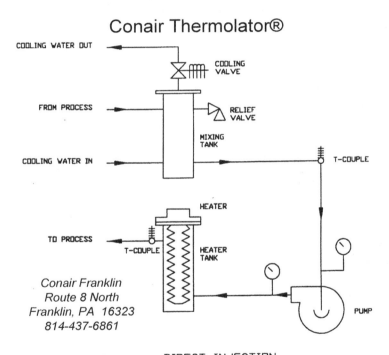

Conair Franklin
Route 8 North
Franklin, PA 16323
814-437-6861

DIRECT INJECTION

Auxiliary Molding Eq - MTCU (dump valves)

Many molders prefer chilled water for two reasons:
1. The MTCU can now deliver hot or cold water – only limited by chilled water temperature;
2. The chilled water is usually cleaner since it circulates thru a mostly closed system.

It is not energy efficient to heat chilled water, but the MTCU should not be adding large amounts of chilled makeup water when running a hot mold. A mold which has a high heat load (resulting from large cavitation and/or fast cycles and/or large parts) may not receive enough cold makeup water; these installations will require good engineering of the central coolant supply system to ensure that pump and pipe sizing do deliver the coolant flow rates as needed. Such applications may perform better without a MTCU assuming the aforementioned engineering provides adequate flow rates.

There are a couple of important points to consider when selecting a MTCU to ensure that it will be sized as needed for the job. A MTCU accomplishes cooling by dumping hot water; as hot water is dumped, colder makeup water enters the mixing tank. The rate of adding colder makeup water is limited by the rate at which hot water is dumped. This dumping is controlled by a solenoid valve. This valve is typically a 1/4 inch valve which can only pass so much water. Larger valves are available as an option if so specified and purchased. The largest valve is typically a 3/4 inch valve which dumps significantly more hot coolant water from the system. See graphics at right for a comparison of dump valve flow rates and pump performance.

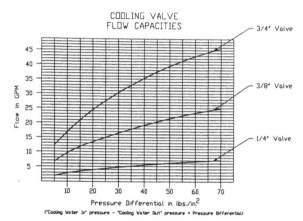

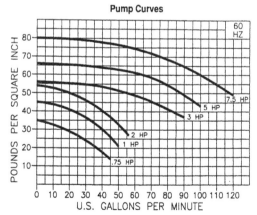

Auxiliary Molding Equipment - Color Feeder

3. COLOR FEEDERS – One style of color feeder is the auger style which is mounted just above the feedthroat and magnet on the molding press. An auger feeder is a volumetric type feeder or measurement and dosing device. This style color feeder permits good colorant level control when sized properly. These feeders typical have digital type set point controls which offer repeatable setups. It is best if the molding press supplies a signal to the color feeder so that the auger run time is simultaneous to the injection molding machine's screw plasticizing.

Proper selection of auger size includes planning or analysis of molding cycle to determine: 1. Time available to plasticize per shot and 2. Shot weight, thus; the color throughput needed to satisfy cycle requirements. For Example: A 45 gram shot weight with 5 seconds of plasticizing time equals 9 grams/sec of blend needed; if the blend ratio is 25:1, we divide 9 by 26 to get 0.346 grams/sec needed ... a 1/2" or 3/4" auger could each work (see also the figure and specifications below). Some people elect to divide by 25 rather than 26 which is only a small difference.

Different sized units are available from a variety of suppliers. Multiple additives (color, lube, etc) can be accomplished with multiple augers. For even greater accuracy, a gravimetric type feeder can be used which proportions each added color or additive based on weight. Full gravimetric blenders are also available for controlling all the ingredients going into a blend.

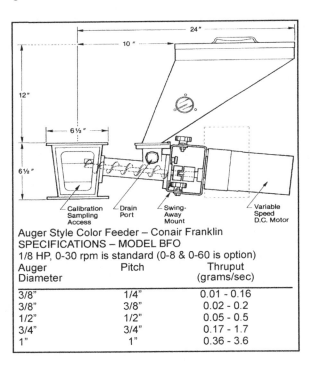

Auger Style Color Feeder – Conair Franklin
SPECIFICATIONS – MODEL BFO
1/8 HP, 0-30 rpm is standard (0-8 & 0-60 is option)

Auger Diameter	Pitch	Thruput (grams/sec)
3/8"	1/4"	0.01 - 0.16
3/8"	3/8"	0.02 - 0.2
1/2"	1/2"	0.05 - 0.5
3/4"	3/4"	0.17 - 1.7
1"	1"	0.36 - 3.6

Deming's 14 Points for Management

1. Create constancy of purpose toward improvement of product and service.

2. Adopt the new philosophy. Acceptance of poor product and service is a roadblock to productivity.

3. Cease dependence on mass inspection. Replace by improved processes.

4. End the practice of awarding business on the basis of price tag alone.

5. Find problems and fix them. Continually reduce waste and improve quality.

6. Institute modern methods of training on the job.

7. Institute modern methods of supervision.

8. Drive out fear.

9. Break down barriers between departments and locations.

10. Eliminate numerical goals, posters and slogans. Don't ask for new levels of productivity without providing methods.

11. Eliminate work standards and numerical quotas.

12. Remove barriers that stand between the worker and his right to pride of workmanship.

13. Institute a vigorous program of education and training.

14. Create a structure in top management that will push every day on the above 13 points.

Decimal Equivalent of Fractions

FRACTION	(IN)	(MM)	FRACTION	(IN)	(MM)
1/64	0.0156	0.3969	33/64	0.5156	13.0969
1/32	0.0313	0.7938	17/32	0.5313	13.4938
3/64	0.0469	1.1906	35/64	0.5469	13.8906
1/16	0.0625	1.5875	9/16	0.5625	14.2875
5/64	0.0781	1.9844	37/64	0.5781	14.6844
3/32	0.0938	2.3813	19/32	0.5938	15.0813
7/64	0.1094	2.7781	39/64	0.6094	15.4781
1/8	0.1250	3.1750	5/8	0.6250	15.8750
9/64	0.1406	3.5719	41/64	0.6406	16.2719
5/32	0.1563	3.9688	21/32	0.6563	16.6688
11/64	0.1719	4.3656	43/64	0.6719	17.0656
3/16	0.1875	4.7625	11/16	0.6875	17.4625
13/64	0.2031	5.1594	45/64	0.7031	17.8594
7/32	0.2188	5.5562	23/32	0.7188	18.2563
15/64	0.2344	5.9531	47/64	0.7344	18.6531
1/4	0.2500	6.3500	3/4	0.7500	19.0500
17/64	0.2656	6.7469	49/64	0.7656	19.4469
9/32	0.2813	7.1438	25/32	0.7813	19.8438
19/64	0.2969	7.5406	51/64	0.7969	20.2406
5/16	0.3125	7.9375	13/16	0.8125	20.6375
21/64	0.3281	8.3344	53/64	0.8281	21.0344
11/32	0.3438	8.7312	27/32	0.8438	21.4313
23/64	0.3594	9.1281	55/64	0.8594	21.8281
3/8	0.3750	9.5250	7/8	0.8750	22.2250
25/64	0.3906	9.9219	57/64	0.8906	22.6219
13/32	0.4063	10.3188	29/32	0.9063	23.0188
27/64	0.4219	10.7156	59/64	0.9219	23.4156
7/16	0.4375	11.1125	15/16	0.9375	23.8125
29/64	0.4531	11.5094	61/64	0.9531	24.2094
15/32	0.4688	11.9063	31/32	0.9688	24.6063
31/64	0.4844	12.3031	63/64	0.9844	25.0031
1/2	0.5000	12.7000	1/1	1.0000	25.4000

1 mm   =   0.03937 inches
1 inch =   25.4 mm

Tap, Number & Letter Drill Sizes

SIZE	T.P.I.	DRILL SIZE	DECIMAL EQUIV	SIZE	T.P.I.	DRILL SIZE	DECIMAL EQUIV	SIZE	T.P.I.	DRILL SIZE	DECIMAL EQUIV	SIZE	T.P.I.	DRILL SIZE	DECIMAL EQUIV
		80	0.0135			37	0.1040			C	0.2420			5/8	0.6250
		79	0.0145		6 32	36	0.1065			D	0.2460		11/16 24	41/64	0.6406
		1/64	0.0156			7/64	0.1094			1/4	0.2500		3/4 10	41/64	0.6406
		78	0.0160			35	0.1100			E	0.2500			21/32	0.6562
		77	0.0180			34	0.1110		5/16 18	F	0.2570		3/4 12	43/64	0.6719
		76	0.0200		6 40	33	0.1130			G	0.2610		3/4 16	11/16	0.6875
		75	0.0210			32	0.1160		5/16 20	17/64	0.2656		3/4 20	45/64	0.7031
		74	0.0225			31	0.1200			H	0.2660			23/32	0.7187
		73	0.0240			1/8	0.1250		5/16 24	I	0.2720		13/16 12	47/64	0.7344
		72	0.0250			30	0.1285			J	0.2770		13/16 16	3/4	0.7500
		71	0.0260		8 32	29	0.1360			K	0.2810		13/16 20	49/64	0.7656
		70	0.0280		8 36	29	0.1360		5/16 32	9/32	0.2812		7/8 9	49/64	0.7656
		69	0.0292			28	0.1405			L	0.2900			25/32	0.7812
		68	0.0310			9/64	0.1406			M	0.2950		7/8 12	51/64	0.7969
		1/32	0.0312			27	0.1440			19/64	0.2969		7/8 14	51/64	0.7969
		67	0.0320			26	0.1470			N	0.3020		7/8 16	13/16	0.8125
		66	0.0330		10 24	25	0.1495		3/8 16	5/16	0.3125		7/8 20	53/64	0.8281
		65	0.0350			24	0.1520			O	0.3160			27/32	0.8437
		64	0.0360			23	0.1540			P	0.3230		15/16 12	55/64	0.8594
		63	0.0370			5/32	0.1562		3/8 20	21/64	0.3281		15/16 16	7/8	0.8750
		62	0.0380		10 32	22	0.1570			Q	0.3320		15/16 20	57/64	0.8906
		61	0.0390			21	0.1590			R	0.3390		1 8	7/8	0.8750
		60	0.0400			20	0.1610		3/8 32	11/32	0.3438		1 12	59/64	0.9219
		59	0.0410			19	0.1660			S	0.3480		1 14	59/64	0.9219
		58	0.0420			18	0.1695			T	0.3580			15/16	0.9375
		57	0.0430		12 24	17	0.1730		7/16 14	U	0.3680		1 16	15/16	0.9375
0 80		56	0.0465			16	0.1770			3/8	0.3750		1 20	61/64	0.9531
		3/64	0.0469		12 28	15	0.1800			V	0.3770			31/32	0.9687
		55	0.0520			14	0.1820		7/16 20	W	0.3860		1-1/8 7	63/64	0.9844
1 64		54	0.0550			13	0.1850			25/64	0.3906			1	1.0000
1 72		53	0.0595			3/16	0.1875		7/16 24	X	0.3970		1-1/8 12	1-1/32	1.0312
		1/16	0.0625			12	0.1890		7/16 28	Y	0.4040			1-3/64	1.0469
		52	0.0635			11	0.1910			13/32	0.4062		1-1/8 16	1-1/16	1.0625
2 56		51	0.0670			10	0.1935			Z	0.4130		1-1/8 18	1-1/16	1.0625
2 64		50	0.0700			9	0.1960		1/2 13	27/64	0.4219				
		49	0.0730			8	0.1990			7/16	0.4375	**PIPE DRILL SIZES (NPT)***			
		48	0.0760		1/4 20	7	0.2010		1/2 20	29/64	0.4531		1/16	D	0.2460
3 48		5/64	0.0781			13/64	0.2031		1/2 24	29/64	0.4531		1/8 27	Q	0.3320
		47	0.0785			6	0.2040		1/2 28	15/32	0.4688		1/4 18	7/16	0.4375
3 56		46	0.0810			5	0.2055		9/16 12	15/32	0.4688		3/8 18	9/16	0.5625
		45	0.0820		1/4 24	4	0.2090			31/64	0.4844		1/2 14	45/64	0.7031
		44	0.0860			3	0.2130		9/16 18	1/2	0.5000		3/4 14	29/32	0.9062
4 40		43	0.0890		1/4 28	7/32	0.2188		9/16 24	33/64	0.5156		1 11½	1-9/64	1.1406
4 48		42	0.0935		1/4 32	7/32	0.2188		5/8 11	17/32	0.5312		1¼ 11½	1-31/64	1.4844
		3/32	0.0938			2	0.2210		5/8 12	35/64	0.5469		1½ 11½	1-47/64	1.7344
		41	0.0960			1	0.2280		5/8 18	9/16	0.5625		2 11½	2-13/16	2.2031
		40	0.0980			A	0.2340		5/8 24	37/64	0.5781				
5 40		39	0.0995			15/64	0.2344			19/32	0.5937	*FOR TAPPING WITHOUT REAMING			
5 44		38	0.1015			B	0.2380		11/16 12	39/64	0.6094				

Right Triangles - Find Angle

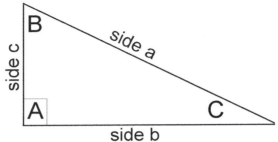

Find Angle:	Formula	
C	$\dfrac{\text{side c}}{\text{side a}}$ =	Sine C
C	$\dfrac{\text{side b}}{\text{side a}}$ =	Cosine C
C	$\dfrac{\text{side c}}{\text{side b}}$ =	Tan C
C	$\dfrac{\text{side b}}{\text{side c}}$ =	Cotan C
C	$\dfrac{\text{side a}}{\text{side b}}$ =	Secant C
C	$\dfrac{\text{side a}}{\text{side c}}$ =	Cosec C
B	$\dfrac{\text{side b}}{\text{side a}}$ =	Sine B
B	$\dfrac{\text{side c}}{\text{side a}}$ =	Cosine B
B	$\dfrac{\text{side b}}{\text{side c}}$ =	Tan B
B	$\dfrac{\text{side c}}{\text{side b}}$ =	Cotan B
B	$\dfrac{\text{side a}}{\text{side c}}$ =	Secant B
B	$\dfrac{\text{side a}}{\text{side b}}$ =	Cosec B

Right Triangles - Find Sides

Find Sides	Formulas	
a	$\sqrt{b^2 + c^2}$	
a	side c X Cosec C	$\dfrac{\text{side c}}{\text{Sine C}}$
a	side c X Secan B	$\dfrac{\text{side c}}{\text{Cosine B}}$
a	side b X Cosec B	$\dfrac{\text{side b}}{\text{Sine B}}$
a	side b X Secan C	$\dfrac{\text{side b}}{\text{Cosine C}}$
b	$\sqrt{a^2 - c^2}$	
b	side a X Sine B	$\dfrac{\text{side a}}{\text{Cosec B}}$
b	side a X Cosine C	$\dfrac{\text{side a}}{\text{Secan C}}$
b	side c X Tan B	$\dfrac{\text{side c}}{\text{Cotan B}}$
b	side c X Cotan C	$\dfrac{\text{side c}}{\text{Tan C}}$
c	$\sqrt{a^2 - b^2}$	
c	side a X Cosine B	$\dfrac{\text{side a}}{\text{Secan B}}$
c	side a X Sine C	$\dfrac{\text{side a}}{\text{Cosec C}}$
c	side b X Cotan B	$\dfrac{\text{side b}}{\text{Tan B}}$
c	side b X Tan C	$\dfrac{\text{side b}}{\text{Cotan C}}$

Conversion Factors
duplicated from page 42

LENGTH
1 inch = 25.4 mm
1 mm = 0.03937 in
1 foot = 30.48 cm
1 micron = 0.001 mm
1 micron = 0.0000394 in
1 inch = 2.54 cm
1 meter = 39.37 in
1 meter = 100 cm
1 microinch = 0.000001 in
1 microinch = 0.0254 microns
(printer's)
1 pica = 0.166 in
1 point = 0.01384 in

WEIGHT
1 lb = 453.6 gr
1 lb = 16 oz
1 gram = 0.035 oz
1 kg = 1000 gr
1 kg = 2.2046 lb
1 oz = 28.35 gr
1 metric ton = 2204.6 lb
1 metric ton = 1000 kg

ANGLES
1 degree = 0.01745 radian
1 degree = $\pi/180$ radian

VOLUME
1 cu in = 16.387 cc
1 cu ft = 1728 cu in
1 qt = 0.946 L
1 gal = 128 oz
1 cc = 1 gr (water)
1 gal = 8.33 lb
1 cu ft = 7.48 gal

AREA
1 sq in = 6.452 cc
1 sq ft = 144 sq in
1 acre = 43560 sq ft
1 sq cm = 0.155 sq in
1 sq ft = 0.111 sq yd
1 sq mm = 0.00155 sq in
1 sq km = 0.3861 sq mi

PRESSURE
1 in Hg = 13.6 in H_2O
1 kg/cm^2 = 14.223 psi
1 bar = 14.5 psi
1 atmos = 14.696 psi
1 MPa = 145 psi

ENERGY
1 BTU = 777.97 ft lb
1 cal = 3.09 ft lb
1 BTU = 252 cal
1 kwh = 3412 BTU
1 H.P. = 746 watts
1 ton (refrig) = 12000 Btu/hr
1 ton (refrig) = 3517 watts

SPECIFIC HEAT & HEAT TRANSFER
1 Cal/sec cm °C = 2903 BTU-in/hr ft^2 °F
1 Cal/sec cm °C = 241.9 BTU/hr ft °F
1 W/(m °K) = 0.0023884 Cal/sec cm °C
1 W/(m °K) = 0.5778 BTU/hr ft °F
1 W/(m °K) = 6.9335 BTU-in/hr ft^2 °F
1 Btu/(Lb °F) = 4.184 KJ/(Kg °K)
1 Btu/(Lb °F) = 4184 J/(Kg °K)
1 Cal/(g °C) = 1 Btu/(Lb °F)

TEMPERATURE CONVERSIONS
°C = (°F-32)/1.8
°F = (°C x 1.8) + 32
°K = (°F+459.67)/1.8

Made in the USA
Monee, IL
30 May 2023

34945462R00059